GROUPS
and
CHARACTERS

GROUPS
and
CHARACTERS

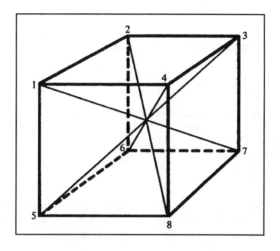

Victor E. Hill IV

Thomas T. Read
Professor of Mathematics
Williams College
Massachusetts, USA

CHAPMAN & HALL/CRC

Boca Raton   London   New York   Washington, D.C.

Library of Congress Cataloging-in-Publication Data

Hill, Victor E., IV
     Groups and characters / Victor E. Hill IV
         p. cm.
     Includes bibliographical references and index.
     ISBN 1-58488-038-4 (alk. paper)
     1. Group theory. 2. Characters of groups. I. Title.
QA174.2.H55 1999
512′ .2—dc21   99-37561
                                                                   CIP

No claim to original U.S. Government works
International Standard Book Number 1-58488-038-4
Library of Congress Card Number 99-37561
Printed in the United States of America   1 2 3 4 5 6 7 8 9 0
Printed on acid-free paper

to my children
Victoria and Christopher

Contents

CONTENTS

Preface

This book is addressed to three potential audiences: (1) undergraduate majors or beginning graduate students in mathematics who would like to have a rapid survey of the theory of groups, representations, and characters at a more modest level of sophistication than that of the classic works by Curtis and Reiner; (2) students in chemistry, physics, or geology, who are likely to encounter groups and characters in such areas as crystallography or quantum mechanics; and (3) physical scientists whose experience with groups and characters, although extensive, has not been rigorous and who would like to have some sense of the mathematics behind the techniques used in applications.

All three of these constituencies have been represented in the courses I have taught for some 30 years at Williams College. In addition, I have had the pleasure of having students who took the group representation theory course become sufficiently interested that they have gone on to more broadbased courses in abstract algebra, moving from groups into rings and fields. This material has also been useful as a supplement to a standard junior/senior course in abstract algebra.

The text assumes a semester of college-level linear algebra and basic familiarity with the style of mathematics exposition. Most of the needed results on matrices and vector spaces are recalled in the text, although only a few are proved here.

This text is intended to be a rapid survey; therefore, it emphasizes examples and applications of the theorems and avoids many of the longer and more difficult proofs (such as those of the orthogonality relations for characters). It also takes for granted basic properties of the natural numbers. Representations are limited to the real and complex fields and, for the most part, to finite groups; for students with some familiarity with finite fields, I have incorporated pertinent examples into my classes, although not into this book.

Some of the exercises stress computation of illustrative examples; my experience has been that students (and other readers) need to spend some time "getting their hands dirty" as they deal with concrete examples. Other exercises stress development of the mathematical theory. The reader or instructor can make an appropriate choice among those provided.

Because this book is an introduction to the mathematical concepts of group and character theory, I have made occasional references to the physical science applications but have not undertaken extended expositions of chemistry or physics; the physical scientists will find plentiful discussions in their own literature.

In the sections on representations and characters, I have adopted for the most part the notations used in the classic works of Curtis and Reiner; they are clear, and their use should facilitate the transition from this introduction to their more comprehensive treatments of the subject. If the reader who uses this work (as an undergraduate text, as supplementary material in an abstract algebra course, or as an exploration of the mathematical side of groups and characters) is thereby motivated to go on to advanced books on representations and characters or other areas of abstract algebra, the purpose served will not be unlike that of a translation of a literary work that leads the reader to learn the language in order to read the text in the original.

My first acknowledgment is to Professor Charles W. Curtis, who introduced me to representation theory, guided my graduate work, and has been a valued friend ever since. The idea of a group acting on a point set was presented in graduate courses by Professor Helmut Wielandt. My students at Williams have used the text in manuscript and have offered numerous helpful suggestions (in addition to catching some annoying typos).

In particular, I am indebted to Stephanie Harding, my editor at Chapman & Hall, and to Helena Redshaw, Evelyn Meany, and Mimi Williams. They have provided the patient and conscientious support for which an author fondly hopes.

Chapter 1

Introductory Examples

To introduce some of the fundamental concepts of group theory in a concrete context, we'll begin with some examples. The terminology introduced informally in this section will be given precise definition in later sections. The object here is to gain a sense of what one thinks about in group theory.

1.1   Example: A matrix group

Consider the matrices:

$$A = \begin{bmatrix} 1 & 0 \\ 1 & -1 \end{bmatrix} \quad \text{and} \quad B = \begin{bmatrix} 0 & -1 \\ -1 & 0 \end{bmatrix}.$$

Here we find

$$A^2 = B^2 = I_2 = \begin{bmatrix} 1 & 0 \\ 0 & 1 \end{bmatrix},$$

the 2×2 identity matrix. We'll say that A and B have *order 2* since the square of each is the identity of the operation under consideration, that of matrix multiplication. Now compute products:

$$AB = \begin{bmatrix} 0 & -1 \\ 1 & -1 \end{bmatrix}, \quad BA = \begin{bmatrix} -1 & 1 \\ -1 & 0 \end{bmatrix};$$

you can easily check that $(AB)^2 = BA$ and that $(AB)^3 = I_2$. Since the cube of AB is the identity matrix I_2, we'll say that AB has *order 3*. You could compute $(BA)^3$ directly or, alternatively, observe that

$$(BA)^3 = ((AB)^2)^3 = (AB)^6 = (I_2)^2 = I_2;$$

either way we see that BA also has order 3.

So far we have found five matrices (including the identity) using the A and B with which we started. Are others possible? At least

$$ABA = \begin{bmatrix} -1 & 1 \\ 0 & 1 \end{bmatrix}$$

can be added to the list since it has not yet been accounted for. To see whether or not others are needed, let's tabulate what we have so far in the form of a multiplication table. We will be able to fill in some of the entries by algebraic computations such as

$$A(ABA) = A^2(BA) = I_2(BA) = BA,$$
$$(ABA)B = (AB)^2 = I_2;$$

others, such as $B(AB)$, can be found by matrix multiplication, where it turns out that $B(AB) = ABA$, a matrix already on our list.

	I_2	A	B	AB	BA	ABA
I_2	I_2	A	B	AB	BA	ABA
A	A	I_2	AB	B	ABA	BA
B	B	BA	I_2	ABA	A	AB
AB	AB	ABA	A	BA	I_2	B
BA	BA	B	ABA	I_2	AB	A
ABA	ABA	AB	BA	A	B	I_2

Note that we were able to complete an entire table without adding any matrices to our list of six. We'll say that the set

$$\{I_2, A, B, AB, BA, ABA\}$$

together with the operation of matrix multiplication forms a *group*, which we may denote by G. Here the letter G is understood to refer to the set of six matrices together with the operation of matrix multiplication. Since the set has six elements, G is called a group of *order 6*.

We have used the term *order* both to designate the number of elements in a group and to specify the smallest positive power of a group element that equals the identity. These two usages will be connected in Proposition 2.11.

1.2   Example: A symmetry group

Consider the solid shown in Figure 1.1; it is called a *trigonal bipyramid*. The vertices have been numbered from 1 to 5. Now we move this solid in space in such a way that the geometric appearance of the solid in space is retained, but the vertices may have been interchanged. For example, if we rotate the solid through 120° about a vertical axis through vertices 1 and 5, those two vertices remain where they were, but 2 goes to the place formerly occupied by 3, 3 to that of 4, and 4 back to that of 2. (Briefly, we'll say, "2 goes to 3," etc.) Let's call this rotation r.

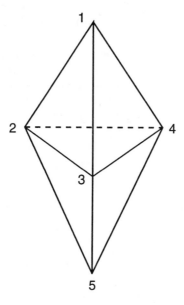

FIGURE 1.1
Trigonal bipyramid

Now if we carry out the rotation r twice, we'll have rotated the solid through 240°, and it is reasonable to write this second rotation as r^2. Note that r performed three times is a rotation through 360°, which is indistinguishable from no movement at all—a sort of identity for the operation of moving the solid in space. We'll denote that identity by e and consequently write

$$r^3 = e.$$

There is a standard notation for such a motion that uses the vertices to specify what has happened. We write

$$r = (234)$$

to mean that vertex 2 has gone to 3, 3 to 4, and 4 back to 2. Since 1 and 5 were left unchanged, we do not ordinarily include them in the expression, but if we wish to emphasize the fact that they have been fixed, we can write

$$r = (1)(234)(5).$$

Similarly, we can write $r^2 = (243)$. Notice that the numbers 2, 3, 4 "cycle around" in r (as 2, 4, 3 do in r^2); in fact, we call a motion such as (234) a *cycle*. If we wrote r as (342) or r^2 as (432), we would express the same motions because these alternative forms would say the same thing about how the vertices move.

Now although r and r^2 left vertices 1 and 5 fixed, it is possible to interchange these two, say, by rotating the solid through 180° about an axis through vertex 4 and the midpoint of the edge joining 2 and 3; the geometrical appearance of the solid would again be maintained even though some vertices have been interchanged. We might call such a motion c; in our notation we could then write

$$c = (15)(23),$$

indicating that vertices 1 and 5 are interchanged, as are 2 and 3. As before, we could write

$$c = (15)(23)(4)$$

to emphasize the fact that vertex 4 is fixed (but accounted for); such emphasis is rarely necessary in practice.

Here we may observe that c^2, meaning c followed by c, is again a rotation through 360°, and again write $c^2 = e$. Moreover, we'll say that c is the *product* of the two cycles (15) and (23).

Obviously, two other 180° rotations are possible: one with the axis through vertex 3 and the other with the axis through vertex 2. Let's see what happens if we perform first r, then c. Here r will fix 1, but c will move it to 5. Similarly, r takes 2 to 3, and then c takes 3 to 2. Checking the rest of the vertices, we see that r followed by c takes 3 to 4, 4 to 3, and 5 to 1. We may then write this "product" of two motions as

$$rc = (15)(34);$$

thus, in fact, rc is the 180° rotation about an axis through vertex 2.

We could also have found rc by a more algebraic method, first writing

$$rc = (234)(15)(23).$$

Now read across the cycles, recapitulating the prose in the preceding paragraph.

1 goes to 5;

2 goes to 3 and then 3 goes to 2;

3 goes to 4;

4 goes to 2 and then 2 goes to 3;

5 goes to 1.

In summary, $rc = (15)(34)$. A similar computation yields

$$r^2c = (15)(24).$$

Note then that $(rc)^2 = e$ and $(r^2c)^2 = e$, which makes sense since rc and r^2c are each 180° rotations. Once again, we'll make a table in which we show the result of two operations, the row corresponding to the first and the column to the second. Instead of listing the elements in the table in the order in which we introduced them, we'll rearrange them for reasons to be seen shortly. As in Example 1.1, we can handle some products algebraically and others by direct computation. For example, $r(rc) = r^2c = r$ and $(rc)c = rc^2 = r$ without direct computation;

$$(r^2c)r = (15)(24)(234) = (15)(34) = rc$$

by a computation that was used above to find the product rc from r and c.

	e	c	rc	r^2	r	r^2c
e	e	c	rc	r^2	r	r^2c
c	c	e	r^2	rc	r^2c	r
rc	rc	r	e	r^2c	c	r^2
r^2	r^2	r^2c	c	r	e	rc
r	r	rc	r^2c	e	r^2	c
r^2c	r^2c	r^2	r	c	rc	e

Again, we were able to complete the table without introducing any further symbols (motions); we'll say that the set $\{e, c, rc, r, r^2, r^2c\}$ together with the operation of successive motions, denoted here as though they were multiplications, forms a group, which we'll call H. Here H, like G, has *order 6*.

Now a comparison of the two tables yields an important observation. If we make the identifications

$$
\begin{aligned}
I_2 &\leftrightarrow e \\
A &\leftrightarrow c \\
B &\leftrightarrow rc \\
AB &\leftrightarrow r^2 \\
BA &\leftrightarrow r \\
ABA &\leftrightarrow r^2 c
\end{aligned}
$$

we find that the tables are identical except for the symbols used to denote the individual elements of G and H. In terms of their mathematical structure, as shown in these "product" tables, G and H are actually the same group. Later we will make this idea explicit in the concept of *isomorphism* and will say that G and H are *isomorphic* and can be called the same *abstract group*.

We have noted that G and H are both of order 6. It is natural to inquire whether or not any other groups have order 6, that is, whether a group of order 6 could have a table that is not the same (except for the symbols chosen to represent the elements) as the one for G and H.

1.3   Example: Another symmetry group

Since we have dealt, in H, with rotations, it should be apparent that we could choose a counterclockwise rotation R through 60° about some axis, say, rotating a hexagon in the plane, and that the table for the rotations induced by R would be

	I	R	R^2	R^3	R^4	R^5
I	I	R	R^2	R^3	R^4	R^5
R	R	R^2	R^3	R^4	R^5	I
R^2	R^2	R^3	R^4	R^5	I	R
R^3	R^3	R^4	R^5	I	R	R^2
R^4	R^4	R^5	I	R	R^2	R^3
R^5	R^5	I	R	R^2	R^3	R^4

where we interpret R^2 as a rotation through 120° (again, counterclockwise), R^3 as one through 180°, etc., until we reach $I = R^6$, a rotation through 360°, which is geometrically the same as a rotation through 0 and is consequently the identity. (Note, by the way, that R^5, a counterclockwise rotation through 300°, is indistinguishable from a clockwise rotation through 60°.) This group is called the *cyclic group of order 6*, and is customarily denoted by $\mathbf{Z}_6$. To see that $\mathbf{Z}_6$ is not the same as G or H above, we need only note that R has order 6 (the smallest power of R that gives the identity I is 6), whereas no element of G or H has order 6.

Now we have yet another question: Are G (or equivalently, H) and $\mathbf{Z}_6$ the *only* possible abstract groups of order 6? The affirmative answer will come in Chapter 8: any group of order 6 is isomorphic to either G or $\mathbf{Z}_6$ in the sense that it has the same table except for the symbols chosen to represent the elements.

The ideas introduced by these three examples will form the basis for our more rigorous discussion in Chapters 2 through 8. In this book we'll also consider groups that arise in chemistry; Exercise 1.8 provides an example.

Exercises

1.1.   Suppose we number the vertices of a regular hexagon from 1 to 6, proceeding counterclockwise around consecutive vertices. Observe that R in the text may be represented as (123456), and then calculate the powers $R^2, \ldots, R^6$.

1.2.   If $p = (123)(45)$ and $q = (135)$, calculate the powers of p and of q, and find the products pq, qp, p^2q, p^3q, and q^2p. (*Note:* The elements p and q here do not represent motions of the solid in Figure 1.1 or, for that matter, of a rigid pentagon in the plane. You might, however, think of them as exchanges among the vertices of a regular pentagon in the plane.)

1.3.   Consider the matrices:

$$C = \begin{bmatrix} 1 & -1 \\ 1 & 0 \end{bmatrix} \quad \text{and} \quad D = \begin{bmatrix} 0 & 1 \\ 1 & 0 \end{bmatrix}.$$

   a.   Show that C, together with its own powers, gives a group having the same table as that for R in Example 1.3.

   b.   In fact, C and D taken together lead to a group of order 12. Find the 12 matrices.

1.4.   Consider the motions of a square in the plane that can be obtained from a rotation r through 90° *clockwise* about the center and a rotation c through $180°$ about an axis joining the midpoints of the two vertical sides (see Figure 1.2).

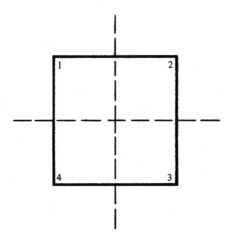

FIGURE 1.2
Square

    a. Observe that $r = (1234)$ and $c = (14)(23)$.

    b. Show that r and c lead to a group of order 8 and make a table for it. This group is called the *dihedral group* of order 8; it will be used extensively in our examples.

1.5. Let

$$R = \begin{bmatrix} 0 & -1 \\ 1 & 0 \end{bmatrix} \qquad \text{and} \qquad C = \begin{bmatrix} 0 & 1 \\ 1 & 0 \end{bmatrix}.$$

    a. Show that $R^4 = I_2$, that $C^2 = I_2$, and that $CR = R^{-1}C$.

    b. Show that R and C lead to a group of order 8 having the same table as that in Exercise 1.4.

1.6. Let

$$A = \begin{bmatrix} 0 & 1 \\ 1 & 0 \end{bmatrix} \qquad \text{and} \qquad B = \begin{bmatrix} -1 & 0 \\ 0 & -1 \end{bmatrix}.$$

Show that A and B lead to a group of order 4.

1.7. Let $i = \sqrt{-1}$ and

$$X = \begin{bmatrix} 0 & i \\ i & 0 \end{bmatrix}.$$

Show that the powers of X give a group of order 4 different from the one found in Exercise 1.6.

1.8. Consider H_2O, a nonlinear molecule, as shown in Figure 1.3. We'll define the following operations, using the standard notation from chemical applications:

σ a reflection in the zy plane,
σ' a reflection in the zx plane,
C_2 a 180° rotation around the z axis.

These operations, together with the identity E, form a group of order 4, which chemists customarily denote as C_{2v}. Make a table for this group. Compare your table to those found in Exercises 1.6 and 1.7; to which of those two groups is C_{2v} isomorphic (in the sense of having the same table except for the symbols used)?

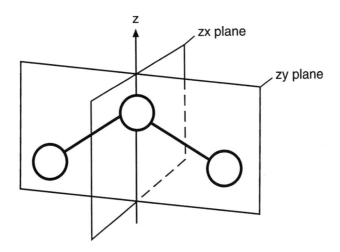

FIGURE 1.3
The H_2O molecule

Chapter 2

Groups and Subgroups

Many mathematical concepts, such as that of a group, are generalized from a variety of specific examples; the three examples presented in the text of Chapter 1, together with those in Exercises 1.3 through 1.8, illustrate this principle. The reason that we are interested in studying groups is that a great many apparently diverse examples have certain features in common that, when abstracted, allow us to study all or some of the examples at once in a general theory. Moreover, group theory has a variety of applications in the physical sciences, which one may explore by using knowledge of the general theory.

The examples in Chapter 1 had some common features. (1) In each case, we combined two elements (matrices or motions) to get another element of the same type. (2) Both the identity matrix and the identity motion (a rotation through 0° or a reflection performed twice), when combined with another element, left that element intact. (3) Each element had an *inverse*, one which, when combined with the original, returned the identity. (4) A property that we used tacitly but crucially was *associativity*, the assumption that

$$(AB)C = A(BC),$$

regardless of whether A, B, and C were matrices or motions in space or in the plane. From such examples, we have the characterization of a group.

2.1   Definition: Group

A *group* G consists of a set, also denoted by G, together with an operation ("multiplication") on pairs of elements of G satisfying the following properties:

(a) If $x, y \in G$, then there is a unique "product," denoted xy, such that $xy \in G$;

(b) If $x, y, z \in G$, then $(xy)z = x(yz)$;

(c) There is an element $e \in G$ such that $ex = xe = x$ for every $x \in G$;

(d) For each $x \in G$, there is an element in G, denoted x^{-1}, for which

$$xx^{-1} = x^{-1}x = e.$$

Property (a) is referred to as *closure*, (b) is the property of *associativity*, the element e in (c) is called the *identity*, and x^{-1} is called the *inverse* of x. Property (c) ensures that G is a nonempty set. In any group G there is only one identity, and for each $x \in G$, there is only one inverse; Exercises 2.10 and 2.11 are concerned with these properties of uniqueness.

The notation in (a), using xy as the product of x and y is accurate, but here xy is just another element of G, equal in status to x and y. We'll most often write a group using the minimal number of letters, just for convenience. However, Exercise 1.8, in which each element had its own symbol, better illustrates the fact that all elements have equal status; that is, a product like xy does not depend upon the elements x and y for its existence as an element in its own right.

One common feature of the examples in Chapter 1 that we did not include in our definition of a group was that of having only a finite number of elements. Although this book is primarily concerned with finite groups, the assumption of finiteness would be too restrictive to take as a part of the definition since many important examples that satisfy Definition 2.1 are infinite.

2.2   Examples: Some infinite groups

(i) The nonzero real numbers with the operation of multiplication form a group.

(ii) The integers with the operation of addition form a group; here the "product" referred to in Property 2.1a is merely the usual *sum*, and the identity element is 0. The inverse of any integer is, naturally, its negative.

(iii) The rational numbers (quotients of the form a/b, where a and b are integers and $b \neq 0$) form a group under addition with 0 as the identity, and the nonzero rational numbers form a group under multiplication with 1 as identity. These may be denoted by $\{\mathbf{Q}, +\}$ and $\mathbf{Q}^*$, respectively.

(iv)   If G consists of a symbol x, together with all of its "powers," x^n, where n is any integer (and x^0 is the identity, denoted by 1), then G forms a group, which we shall call *infinite cyclic* and denote by $\mathbf{Z}_\infty$. Note that $(x^n)^{-1} = (x^{-1})^n$ for any n (see Exercise 2.12).

(v)   Let $\mathbf{R}^2$ denote the usual xy-plane, k be an irrational number, and R be a counterclockwise rotation of $\mathbf{R}^2$ through angle $\theta = 2\pi/k$ about the origin. For a positive integer c, R^c denotes a counterclockwise rotation through angle $c\theta$, for a negative integer c; R^c denotes a clockwise rotation through angle $c\theta$; and R^0 is the identity. Then all of these powers of R form another infinite cyclic group, distinguished from Example iv only in that R was used instead of x.

(vi)   Let $(\mathbf{Q})_n$ denote the $n \times n$ nonsingular (i.e., invertible) matrices with entries from the rational numbers $\mathbf{Q}$. Then $(\mathbf{Q})_n$, with the operation of matrix multiplication, forms a group with the identity matrix I_2 as identity.

(viii)   Let S be the sphere in three-dimensional space with the center at the origin and radius 1. Then the set of all rotations of S about the origin forms a noncyclic infinite group.

Note that we have allowed a *sum* as the group operation, using 0 as the identity; when we do so, we say that we are writing the group *additively*, as distinguished from *multiplicatively*, where the product notation uses 1 as the identity. In a symmetry group like those in Examples 1.1 and 1.2, we typically use multiplicative notation; the choice of a symbol for the identity (i in Example 1.1 and I in Example 1.2) is made to avoid ambiguity, here with vertex 1. Context always plays a role in a choice of notation; for example, we would certainly not want to use i for the identity when we were also using it to mean $\sqrt{-1}$.

Let's make explicit, for finite groups, a term used uniformly in Chapter 1.

2.3   Definition: Order of a finite group

The *order* of a group G having a finite number of elements is that number, the order of G is noted by $|G|$.

We'll occasionally use the notation $|S|$ also the number of elements in a finite set S, even if that set is not endowed with a group operation.

Two important observations about the inverse are as follows.

2.4   Proposition: Inverse of the inverse

If $g \in G$, then $(g^{-1})^{-1} = g$.

Proof: This observation is apparent from Property 2.1d.   □

2.5   Proposition: Inverse of a product

If $x, y \in G$, then $(xy)^{-1} = y^{-1}x^{-1}$.

Proof: Using Property 2.1b,

$$(y^{-1}x^{-1})(xy) = y^{-1}(x^{-1}x)y = y^{-1} 1\, y = y^{-1}y = 1, \qquad \text{and}$$

$$(xy)(y^{-1}x^{-1}) = x(yy^{-1})x^{-1} = x\, 1\, x^{-1} = xx^{-1} = 1. \qquad □$$

Note that we did not specify in Proposition 2.4 or 2.5 that G is a group. Let's just make the convention that, from now on, G *always* refers to a group.

Exercise 1.6 provided an important example of a group of order 4, which in the abstract is called the *Klein 4-group* and denoted by V_4. (This notation is, in fact, redundant since the V stands for the German *vier*, meaning *four*, but it is customary.) This group may be written as the set $\{1, a, b, ab\}$ with the table

	1	a	b	ab
1	1	a	b	ab
a	a	1	ab	b
b	b	ab	1	a
ab	ab	b	a	1

Note that V_4 has a property that it shares with Example 1.3 but not with the other two examples in Chapter 1, namely, that $xy = yx$ for every choice of x and y in the group. We did not make this property part of Definition 2.1 because there are many important examples of groups in which it fails to hold. However, we do distinguish those groups in which it does hold.

2.6 Definition: Abelian or commutative group

A group G is called *abelian* or *commutative* if

$$xy = yx \qquad \text{for all } x, y \in G.$$

The term *abelian* honors Niels Henrik Abel (1802–1829), one of the first mathematicians to work in what we now know as group theory. Abelian groups form an important special case, and they will play a central role in our construction of character tables in later sections of this book. You probably met the antonym *noncommutative* in your study of linear algebra, where it describes matrix multiplication; we'll also use the term *nonabelian* for a group that does not have the commutative property.

Looking at the table for V_4, we note that the subset $\{1, a\}$ determines the limited table

	1	a
1	1	a
a	a	1

and thus that this subset forms a (smaller) group in its own right. The same can be said for the subsets $\{1, b\}$ and $\{1, ab\}$. (It cannot be said of the subsets $\{1, a, b\}$ and $\{a, b\}$. Which of the group axioms fail to hold here?) This leads us to the following definition.

2.7 Definition: Subgroup

A nonempty subset H of a group G is a *subgroup* if H satisfies the conditions in Definition 2.1. (The operation in H is taken to be the same as that in G, but applied only to the elements of H.)

If 1 is the identity of G, then it is clear that $\{1\}$ is a subgroup, and is called the *identity subgroup*. Likewise, G is itself a subgroup of G. On the other hand, the empty set $\emptyset$ is never a subgroup of any group because it violates Condition 2.1c.

From here on, we'll use the notation

$H \subseteq G$ for H is a subset of G,
$H \leq G$ for H is a subgroup of G,
$H \subset G$ for H is a proper subset of G,
$H < G$ for H is a proper subgroup of G,

where, as usual, *proper* means that $H \neq G$.

To show that a subset H of G is a subgroup, we do not need to check the associative property because $(xy)z = x(yz)$ certainly holds in the smaller set H if it holds in the larger one G. Thus, we have the following useful criterion.

2.8   Proposition: Characterization of subgroup

A subset H of a group G is a subgroup if and only if

(a) H is nonempty;

(b)   if $x, y \in H$, then $xy \in H$;

(c)   if $x \in H$, then $x^{-1} \in H$.

In Proposition 2.8, condition b follows immediately from 2.8a and c: if H satisfies conditions a, b, and c, then H has some element x (by a), hence $x^{-1} \in H$ (by c), and therefore $xx^{-1} \in H$ (by b), but $xx^{-1} = e$, so $e \in H$. In applying Proposition 2.8 to a specific subset H, we will most often check condition a by noting that the identity of G is in H; note, conversely, that if a subset H fails to contain the identity element of G, it *cannot* be a subgroup of G.

Sometimes it is convenient to use a more concise form of Proposition 2.8 as follows.

2.9   Proposition: Another characterization of subgroups

A nonempty subset H of G is a subgroup of G if and only if $xy^{-1} \in H$ whenever $x, y \in H$.

Proof: If $H \leq G$, then the conclusion follows immediately from Proposition 2.8. Conversely, let $x \in H$. If the stated condition holds for H, then replace the y by x, which we may do because x and y are merely variables representing arbitrary elements of G; this results in $xx^{-1} = e \in H$. Now apply the condition to e and x: $ex^{-1} = x^{-1} \in H$, as required by Condition 2.8c. Finally, for $x, y \in H$, we know that $y^{-1} \in H$ by the preceding and that $y = (y^{-1})^{-1}$ by Proposition 2.4. Therefore,

$$xy = x(y^{-1})^{-1} \in H$$

by the given condition. $\qquad\square$

An important question about finite groups is what we can say about the existence of subgroups, given the order of the group. A first step in this direction comes from the idea of the order of an element. If $g \in G$, then the set of all distinct powers of g, denoted by $\langle g \rangle$, forms a subgroup of G, called the subgroup *generated by* g. Hence, the definition that follows.

2.10 Definition: Order of an element

If $g \in G$ and if there exists a positive integer n such that $g^n = 1$, then the *smallest* such n is called the *order of* g.

Of course, the order of g is also the order of the subgroup $\langle g \rangle$. We have the following immediately.

2.11 Proposition: Subgroup generated by an element

If $g \in G$ has order n, then G has a subgroup of order n, namely, $\langle g \rangle$.

A group consisting of the distinct powers of a single element of finite order n is called the *cyclic group of order n* and is denoted by $\mathbf{Z}_n$.

We conclude this section with some basic, but important, properties as follows.

2.12 Proposition: Solution of equations

If $a, b \in G$, then there exist unique elements $x, y \in G$ such that

$$ax = b \quad \text{and} \quad ya = b.$$

2.13   Proposition: Cancellation

Let $a, b, c \in G$.

(a)   If $ac = bc$, then $a = b$.

(b)   If $ab = ac$, then $b = c$.

The proofs are left as exercises.

Exercises

2.1.   Find all of the subgroups of the group in Example 1.1. Make a table that shows how these subgroups are contained in one another.

2.2.   Find all of the subgroups of the group in Example 1.2. Make a table that shows how these subgroups are contained in one another.

2.3.   Find all of the subgroups of the group in Example 1.3. Make a table that shows how these subgroups are contained in one another.

2.4.   Show that the only subgroups of V_4 are $\{1\}$, V_4 itself, and those accounted for in the paragraph preceding Definition 2.7.

2.5.   Show that $\mathbf{Z}_\infty$ has infinitely many subgroups.

2.6.   Prove that if $K \le H$ and $H \le G$, then $K \le G$.

2.7.   If $H \le G$ and $K \le G$, prove that $H \cap K \le G$.

2.8.   Show by specific example that, if $H \le G$ and $K \le G$, then $H \cup K$ is not, in general, a subgroup of G.

2.9.   In the table for Example 2̶.2̶, (1.2), $(rc)^{-1} = rc$; explain why this fact does not contradict Proposition 2.5, which says that $(rc)^{-1} = c^{-1}r^{-1} = cr^2$.

2.10.   Show that the identity element of a group is unique; that is, if e and e^* are both elements of a group G for which

$$ex = xe = x \qquad \text{and} \qquad e^*x = xe^* = x \qquad \text{for every } x \in G,$$

then $e = e^*$.

2.11. Show that each element x of a group G has only one inverse; that is,

$$\text{if } xx^{-1} = x^{-1}x = e \qquad \text{and} \qquad xx^* = x^*x = e, \qquad \text{then } x^{-1} = x^*.$$

2.12. Let $g \in G$. Show that $(g^2)^{-1} = (g^{-1})^2$. This result shows that we may write g^{-2} unambiguously and, by extension, g^{-k} for any integer k.

2.13. Prove Proposition 2.12.

2.14. Prove Proposition 2.13.

2.15. Show that every subgroup of a cyclic group is also cyclic.

2.16. Let G be a group and H a subgroup of G. What can you say about the sets

$$\{g^{-1} : g \in G\} \qquad \text{and} \qquad \{h^{-1} : h \in H\}?$$

We'll need to use the answer to this question later.

2.17. Recall the discussion of H_2O in Exercise 1.8, in which we had a group C_{2v} of order 4 having the following set of operations:

$$C_{2v} = \{E, C_2, \sigma, \sigma'\}.$$

Let's define an additional element t to be the translation of the entire molecule one unit along the z-axis in the positive direction.

a. Does the set of five operations $\{E, C_2, \sigma, \sigma', t\}$ form a group?

b. Consider an additional element t^{-1}, a translation one unit along the z-axis in the negative direction; does the set of six operations $\{E, C_2, \sigma, \sigma', t, t^{-1}\}$ form a group?

Chapter 3

Point Groups and Cosets

We may connect the examples of symmetry groups from Chapter 1 and the abstract definition of a group from Chapter 2 by considering a set of "points" Ω on which a group G acts.

3.1   Definition: Action of a group on a point set

We say that a group G *acts on* a set of points Ω if, for each $g \in G$, a correspondence is set up which takes each point $\alpha \in \Omega$ to a (not necessarily distinct) point $\alpha^g \in \Omega$ such that the following conditions are met:

$$(\alpha^g)^h = \alpha^{gh} \qquad \text{for} \qquad g, h \in G;$$
$$\alpha^e = \alpha \qquad \text{for the identity } e \text{ of } G.$$

It is important to realize that the term *point* is used in an abstract fashion here; instead of the vertices in Example 1.2 we might have considered the action of the group G of motions on the faces of the trigonal bipyramid, and instead of thinking in terms of the vertices of the hexagon in Example 1.3, we could have spoken of transformations of the diagonals of that plane figure. Faces or diagonals would still be "points" for the purposes of this definition.

Another note is that α^g is not a "power" of α; the notation is merely a convenient one to designate the point of Ω to which g carries the point α; thus α^g is merely another point of Ω, just as a product xy of group elements x and y is just another element of the group.

We said in the previous section that G would always denote a group; similarly, Ω will always denote a "point set" upon which some G acts.

Some additional vocabulary is needed now.

3.2   Definition: Orbit and stabilizer

If the group G acts on the set Ω and $\alpha \in \Omega$, then the *stabilizer of α in G* is the following subset: $\{g \in G : \alpha^g = \alpha\}$, denoted by G_α. The *orbit of α under G* is $\{\alpha^g : g \in G\}$, denoted by α^G.

Thus the stabilizer of α in G is the set of elements of G that leave α fixed, and the orbit of α under G is the set of points to which elements of G carry α. Note that G_α is a subset of G, whereas α^G is a subset of Ω. The two are not obviously comparable, although we shall see in Theorem 3.13 that an important relationship exists between the orders of these sets.

The idea of subgroup enters here.

3.3   Proposition: The stabilizer is a subgroup

If G acts on Ω and $\alpha \in \Omega$, then $G_\alpha \leq G$.

Proof: By Definition 3.1, the identity of G fixes α, so G_α is nonempty. If $x, y \in G_\alpha$, then $\alpha^{xy} = (\alpha^x)^y = \alpha^y = \alpha$, so $xy \in G_\alpha$. Likewise, if $x \in G_\alpha$, then

$$\alpha = \alpha^{xx^{-1}} = (\alpha^x)^{x^{-1}} = \alpha^{x^{-1}},$$

so $x^{-1} \in G_\alpha$. By Proposition 2.8, $G_\alpha \leq G$. $\square$

Now for fixed $x \in G$, let's consider the subset of G defined by

$$G_\alpha x = \{gx : g \in G_\alpha\}.$$

By our Definition 3.1, x takes the point α to some point β, but then if $g \in G_\alpha$, then gx also takes α to the same β because

$$\alpha^{gx} = (\alpha^g)^x = \alpha^x = \beta.$$

Thus, this set $G_\alpha x$ consists *entirely* of elements of G that carry α to β. Exercise 3.3 asks you to show that no elements of G other than those in $G_\alpha x$ take α to β. Hence, $G_\alpha x$ consists *exclusively* (and thus *precisely*) of those elements of G that carry α to β; that is,

$$G_\alpha x = \{g \in G : \alpha^g = \alpha^x\}.$$

Now suppose that there are exactly n elements in the orbit of the point α. If we had elements $x_1, \ldots, x_n$ of G for which the various α^{x_i} (for $i = 1, \ldots, n$) were precisely those n elements of the orbit, we would have all of the elements of the point group G; that is, we would know that

$$G = G_\alpha x_1 \cup \cdots \cup G_\alpha x_n.$$

Moreover, since an element in $G_\alpha x_1$ carries α to some point β, and since an element in, say, $G_\alpha x_2$, carries α to some *other* element γ, the subsets $G_\alpha x_1$ and $G_\alpha x_2$ must be disjoint. All of this discussion is intended to motivate the following definition and theorem.

3.4 Definition: Coset and coset representative

Let $H \leq G$ and $g \in G$. We define the *right coset of g* to be the set

$$Hg = \{hg : h \in H\}$$

and all g a *representative of this coset.* Since the identity 1 of G is in H, each g is a member of its own coset Hg. Moreover, consider the following.

3.5 Proposition: $H(hg) = Hg$ for cosets

If $H \leq G$ and if $h \in H$, then

$$H(hg) = (Hh)g = Hg.$$

Proof: The first equality holds by associativity and the definition of Hh, and the second because H has closure; that completes the proof. □

Of course, Proposition 3.5 includes the fact that $Hh = H$ whenever $h \in H$. These seemingly simple facts will often be used in our arguments about cosets, including the proof of Theorem 3.6.

It is natural to define a *left coset* in a similar fashion as

$$gH = \{gh : h \in H\},$$

and we'll use left cosets in our discussion of induced representations and characters in Chapter 19. The two types of cosets are related in Exercise 3.8.

Now let's return to the decomposition of a group G acting on a set Ω into right cosets of the form $G_\alpha x$. Having noted that each $G_\alpha x$ consists precisely of those elements of G that carry α to the same point as x does, we observed that two such cosets must either coincide or be disjoint. In the more general context of a subgroup H of a group G, it is clear that every $g \in G$ belongs to the coset Hg. The observation that cosets either coincide or are disjoint in this more general case is shown by our next result.

3.6 Theorem: Cosets coincide or are disjoint

Let $H \leq G$ and $x, y \in G$. Then either $Hx = Hy$ or $Hx \cap Hy = \emptyset$.

Proof: We suppose that $Hx \cap Hy \neq \emptyset$ and show that $Hx = Hy$. If $Hx \cap Hy \neq \emptyset$, then the two cosets have a common element z, and by the definition of coset, there must exist $h, k \in H$ such that $z = hx = ky \in Hx \cap Hy$. But then, multiplying on both sides, we have

$$h^{-1}(hx)y^{-1} = h^{-1}(ky)y^{-1},$$

which simplifies by associativity to $xy^{-1} = h^{-1}k \in H$, and so we conclude that $xy^{-1} \in H$. An arbitrary element of Hx can be written, using some $m \in H$, as

$$mx = (mx)(y^{-1}y) = (m(xy^{-1}))y \in Hy$$

since $xy^{-1} \in H$ and by closure $m(xy^{-1}) \in H$; hence, $Hx \subseteq Hy$. If $ny \in Hy$ with $n \in H$, then

$$ny = (ny)(x^{-1}x) = (n(xy^{-1})^{-1})x \in Hx$$

so $Hy \subseteq Hx$ also. This completes the proof.   □

The question of when two cosets are equal is nicely characterized by the following.

3.7 Proposition: Equality of cosets

If $H \leq G$ and $x, y \in G$, then $Hx = Hy$ if and only if $xy^{-1} \in H$.

Proof: If $Hx = Hy$, then since the identity $e \in H$, and since $x = ex$, the element $x \in Hx$ and consequently, $x \in Hy$ as well. Therefore, there exists $h \in H$ such that $x = hy$ (recall that equality of sets means that the two have exactly the same elements). But then $xy^{-1} = h \in H$. You should verify that the converse was proved along with Theorem 3.6. □

We'll denote the group introduced in Example 1.2 as D_3, where the D means "dihedral" and designates the group of the symmetries of a regular polygon, with the subscript denoting the number of sides; thus D_3 is the group of symmetries of the equilateral triangle. Note that this group has order 6; we also say that D_3 is the *dihedral group of degree 3*. (The group D_4 of the square, which has order 8, appeared in Exercise 1.4.) You should be cautioned, however, that some notations in modern mathematics are not unambiguous and, in particular, that some authors write D_6 for the dihedral group of order 6 (and of degree 3).

Of course, Example 1.2 used a solid figure, but that was so that we could look at orbits and stabilizers of the vertices (see the Exercises for this chapter); a geometric consideration of the solid shows that if you know the position of the central triangle, you know the position of the solid.

3.8 Example: Cosets in the group D_3

In the notation of Example 1.2, $\langle r \rangle = \{e, r, r^2\}$ forms a subgroup, whose cosets are $\langle r \rangle$ itself and $\langle r \rangle c = \{c, rc, r^2c\}$. The subgroup $\langle c \rangle = \{e, c\}$ has cosets $\langle c \rangle$, $\langle c \rangle r = \{c, r^2c\}$, and $\langle c \rangle r^2 = \{r^2, rc\}$; here $r^2c \in \langle c \rangle r$ because $r^2c = cr$.

It is important to note that cosets do not have unique designations in the form Hg. Note that, in the preceding example, $\langle c \rangle r = \langle c \rangle r^2 c$ because

$$\langle c \rangle r^2 c = \{r^2 c, cr^2 c\} = \{r^2 c, c\} = \langle c \rangle r$$

using the table given in Example 1.2.

Since we now know that any two cosets of a group either coincide or are disjoint, we can compare cosets.

3.9   Proposition: Cosets have the same order

If H is a subgroup of the finite group G and if $x, y \in H$, then $|Hx| = |Hy|$.

Proof: If $hx = kx$ for elements $h, k \in G$, then $h = k$ by cancellation, so $|Hx| \geq |H|$. But $|Hx| \leq |H|$ by definition of Hx. Hence $|Hx| = |H| = |Hy|$ for any $y \in G$ also.   □

We are thus led to what is called the theorem of Lagrange. Group theory as such did not yet exist in the time of Jean-Joseph Lagrange (1736–1813), but the essence of the following result may be distilled from his writings, and for that reason the conclusion has come to be known by his name.

3.10   Theorem of Lagrange: Order of a subgroup

If G is a finite group and $H \leq G$, then $|H|$ divides $|G|$.

Proof: G is divided into disjoint cosets of H, each having the order of H.   □

Theorem 3.10 leads to several consequences. First, we look at the idea of the index of a subgroup.

3.11   Definition: Index of a subgroup

The *index* of a subgroup H of a finite group G, denoted by $[G : H]$, is the number of distinct cosets of H in G.

Immediately, we have the following theorem.

3.12   Theorem: Index and order of a subgroup

If H is a subgroup of a finite group G, then

$$|G| = |H|\,[G : H].$$

(The implied operation in this equation is, of course, ordinary multiplication since the terms are positive integers.)

We'll use this result many times in our study of groups. But it happens that a result regarding groups acting on a set is rather similar and, likewise, of importance.

3.13 Theorem: Orders of orbit and stabilizer

If a group G acts on a set Ω and if $\alpha \in \Omega$, then

$$|G| = |\alpha^G||G_\alpha|.$$

Before we take up the proof of this result, let's look at what it says. Since the orbit of α is a subset of Ω and the stabilizer of α is a subset of G, these two "animals" are not part of the same "species," but this result says that their orders, which are positive integers, are closely related to the order of the group of actions on Ω.

Proof: Since $G_\alpha \leq G$, we know by Theorem 3.12 that $|G| = |G_\alpha|[G : G_\alpha]$, so what we need to show is that $|\alpha^G| = [G : G_\alpha]$. Let $x, y \in G$; then

$$\alpha^x = \alpha^y \quad \text{if and only if } \alpha^{xy^{-1}} = \alpha,$$
$$\text{if and only if } xy^{-1} \in G_\alpha,$$
$$\text{if and only if } G_\alpha x = G_\alpha y \text{ by Proposition 3.7.}$$

Thus α^x and α^y are *distinct* precisely when the cosets $G_\alpha x$ and $G_\alpha y$ are distinct; hence the number of points in the orbit α^G is precisely the number of cosets of G_α in G. □

This theorem can be used to find the order of a point group, as shown in the following example.

3.14 Application: The group of the cube

Consider a solid cube; take Ω to consist of the eight vertices, as shown in Figure 3.1, and let G be the symmetry group, as was done for the trigonal bipyramid in Examples 1.2 and 1.3. To find the stabilizer G_1 of vertex 1, consider the three edges meeting at 1. Since vertex 1 is to be fixed, these three edges can only be permuted among themselves; specifically, we have rotations of 120° and 240° about the axis formed by the diagonal joining vertices 1 and 7, and we also have the identity. Thus $|G_1| = 3$, and since $|1^G| = |\Omega| = 8$, Theorem 3.13 tells us that the order of G is 24.

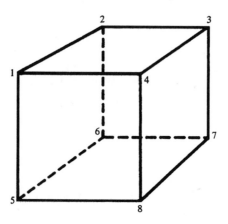

FIGURE 3.1
Cube with numbered vertices

3.15   Application: The "full" cubic group

For applications in the physical sciences, the cube (which might represent a crystal of cesium chloride, $CsCl_8$) is not treated as rigid but can be inverted through its own center; this inversion may be written in the notation of Figure 3.1 as the permutation

$$z = (17)(28)(35)(46).$$

A larger group G^* results and, in fact, $G \leq G^*$ with the coset decomposition

$$G^* = G \cup Gz.$$

Thus $|G^*| = 48$. To perform the operation z on a hollow cube with solid faces, one would have to turn it inside out.

Exercises

3.1. In the group D_3 of Example 1.2, find the orbits and stabilizers of the points 1 and 3.

3.2. Find the right cosets of the subgroup $\langle rc \rangle$ in the group D_3.

3.3. Let G act on Ω, $\alpha \in \Omega$, $x \in G$, and $\alpha^x = \beta$. Show that if $y \in G$ and $\alpha^y = \beta$, then there exists $g \in G_\alpha$ for which $y = gx$; that is, $y \in G_\alpha x$.

3.4. The group introduced in Exercise 1.4 is customarily denoted as D_4 and called the *dihedral group of order 8*. (Compare the group D_3 in Example 3.8.) Find the orbits and stabilizers of the points 1 and 4.

3.5. If $H \leq G$ and $x \notin H$, show that Hx is *not* a subgroup of G.

3.6. The regular tetrahedron is shown in Figure 3.2; it has four vertices. Use Theorem 3.13 to find the order of the symmetry group of the tetrahedron.

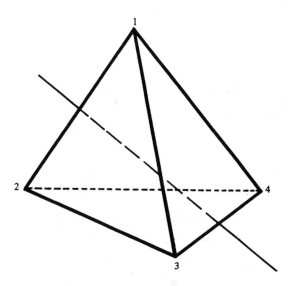

FIGURE 3.2
Tetrahedron

3.7. For the "full" cubic group G^* in Proposition 3.15, find G_1^* and verify that Theorem 3.13 still holds.

3.8.   To see the relationship between right and left cosets of a subgroup H in G, let

$$G = Hg_1 \cup Hg_2 \cup \cdots \cup Hg_n;$$

show that

$$G = (g_1)^{-1}H \cup (g_2)^{-1}H \cup \cdots \cup (g_n)^{-1}H.$$

3.9.   Show by an example using the group D_4 that if

$$G = Hg_1 \cup Hg_2 \cup Hg_3 \cup Hg_4,$$

it need *not* be the case that

$$G = g_1H \cup g_2H \cup g_3H \cup g_4H.$$

Thus the inverses taken in Exercise 3.8 are necessary.

3.10. Prove that a group of prime order must be cyclic.

✶ 3.11. In Exercise 2.8 you were asked to show that if $H \leq G$ and $K \leq G$, then $H \cup K$ is not, in general, a subgroup of G. Now prove that if $H \cup K$ is, in fact, a subgroup of G, then either $H \subseteq K$ or $K \subseteq H$. [*Hint*: Suppose that H is not contained in K; then there is an element $h \in H$ such that $h \notin K$. Show that an arbitrary k in K must be in H also.]

3.12. Refer to the group of the tetrahedron (Exercise 3.6 and Figure 3.2). Now let Ω be the four faces of the solid, say

$$\Omega = \{A, B, C, D\}, \qquad \text{where } A = 234, B = 134, C = 124, D = 123.$$

Then $A^{(123)} = B$, $A^{(132)} = C$, and so forth. (Be careful to distinguish between the permutation (123), a member of the group, and the face $D = \overline{123}$, written without overline.) An element of the stabilizer of A may rotate the face A so long as it does not move A to another face of the tetrahedron. Find the orbits and stabilizers of A and C.

3.13. Again using the group of the tetrahedron, let Ω consist of the edges of the solid. An edge is fixed by a group element if it is turned around, as (12)(34) fixes the edges 12 and 34, just so long as the edge is not moved to a different edge. Find the orbits and stabilizers of the edges 12 and 23 and verify that Theorem 3.13 still holds.

3.14. Let G be the group of the tetrahedron from Exercise 3.6. From chemistry, we know that the methane molecule CH_4 possesses a tetrahedral structure with a hydrogen atom at each vertex and a carbon atom at the center, as shown in Figure 3.3. Let H be the subset of G consisting of the identity and each of the 180° rotations around the lines which bisect opposite edges.

a. Verify that H is a subgroup, and use Theorem 3.13 to find its order.

b. Find the orbits and stabilizers under H of each of the hydrogen atoms.

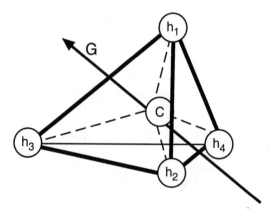

FIGURE 3.3
Methane

Now consider methane, as shown in Figure 3.4. Let F consist of the identity, reflection in the plane $C12$, reflection in the plane $C34$, and a 180° rotation around the z-axis.

c. Use Theorem 3.13 to find the order of F.

d. Find the orbits and stabilizers under F of each of the hydrogen atoms.

e. Make tables for the subgroups H and F, and compare these tables with one another and also with the table from Exercise 1.8.

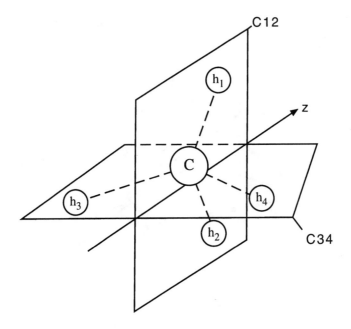

FIGURE 3.4
Methane, rev.

Chapter 4

Homomorphisms and Normal Subgroups

In Chapter 2 we defined a group on the basis of properties we observed in the examples in Chapter 1. The essentials were a set of elements and an operation with some rather limited properties. We also observed that Examples 1.1 and 1.2 were essentially the same, in the sense that the operations had the same table except for the choice of symbols used to represent the elements. The question, then, is how a set with a group operation might be transformed while preserving some or all of the structure. You should recall that in linear algebra you had *linear transformations* that preserved the structure of a vector space; functions analogous to linear transformations are basic to group theory. We'll use the notation $\phi : G \to G^*$ to mean that ϕ is a function from G into G^*.

4.1 Definition: Homomorphism

Let G and G^* be groups; a function $\phi : G \to G^*$ is a *homomorphism* if $\phi(xy) = \phi(x)\phi(y)$ for every choice of x and y in G; ϕ is said to *preserve products*.

Notice that two different operations are involved here: the product xy takes place in G, and the product $\phi(x)\phi(y)$ in G^*. If, for example, we defined ϕ from the group in Example 1.1 to the one in Example 1.2 with $\phi(A) = c$ and $\phi(B) = rc$, we would need to verify that ϕ carried the matrix product AB to the symmetry product $c(rc)$.

Recall that linear transformations preserve vector sums and that they also preserve zero vectors and negatives of vectors. A similar result holds for homomorphisms.

4.2   Proposition: Homomorphisms preserve identity and inverse

Let $\phi : G \to G^*$ be a homomorphism, where e and e^* are the identity elements of G and G^*, respectively. Then $\phi(e) = e^*$ and, if $x \in G$, then

$$\phi(x^{-1}) = \left(\phi(x)\right)^{-1}.$$

(Note again, that the inverse operations take place in different groups: x^{-1} in G, and $\phi(x)^{-1}$ in G^*.)

Proof: For $x \in G$, $\phi(x) = \phi(ex) = \phi(e)\phi(x)$; multiplying both sides on the right by $\phi(x)^{-1}$, we have $e^* = \phi(e)$. For the second part,

$$e^* = \phi(e) = \phi(xx^{-1}) = \phi(x)\phi(x^{-1}),$$

so $\phi(x^{-1}) = \phi(x)^{-1}$ by the uniqueness of inverses in G^* (see Exercise 2.11).   □

Recall that a linear transformation of a vector space is completely determined by its action on the elements of a basis for that space. In the same way, if a group G is written entirely in terms of one of more symbols, which we call *generators*, then a homomorphism is determined by what it does to those generators because products must be preserved, and powers are just special cases of products. Consider the following.

4.3   Example: Homomorphism defined on generators

Let G and G^* be the groups in Examples 1.1 and 1.2, and for $\phi : G \to G^*$, let's specify that $\phi(A) = c$ and $\phi(B) = rc$, as in the discussion above. This more fully determines all of the values of ϕ, if it is a homomorphism at all. We must have $\phi(A^2) = c^2 = i$, $\phi(B^2) = (rc)^2 = i$, and $\phi(BA)$ has to be $(rc)r = c$ if ϕ is to preserve products; likewise, $\phi(AB)$ must be $c(rc) = r^2$, and $\phi(ABA)$ must be $c(rc)c = r^2c$. That all of the possible products are preserved could be verified from the tables, but to avoid such tedium, we may bring in the concept of a group defined *in terms of generators and relations*.

The idea here is quite straightforward, but it is easiest to understand for finite groups, so we'll assume that G has finite order and, hence, so do all of its elements. We have a list of *generators*, and we have some *relations* of equality that are assumed to hold. If that information suffices to determine one and only one group table, we have specified the group in terms of generators and relations.

In Chapter 2 we noted that, for an element g in a finite group G, $\langle g \rangle$ is a subgroup that may be thought of as a cyclic group $\mathbf{Z}_n$, where n is the order of the element g. Thus it is natural to begin our definition with the orders of the generators. Then we add enough relations to determine the group table once and for all. In Example 1.2, we started with motions r and c but listed the elements as e, c, rc, etc. to show the connection with Example 1.1. If we go back and specify that

$$r^3 = e, \qquad c^2 = e, \qquad cr = r^2 c,$$

you can verify that these relations tell you all you need to know to complete the table. To show that the group is generated by r and c, with these three relations, we'll use the notation

$$G = \langle r, c : r^3 = 1, c^2 = 1, cr = r^2 c \rangle$$

where we have used 1 in place of e for the identity.

Two cautions are needed here. The first is that when we write an expression like $r^3 = 1$ in this context, we mean that the element r has order 3, and nothing less. Saying $r^k = 1$ in this context means that $r^j \neq 1$ for $1 \leq j < k$. The second is that writing down a set of generators and relations at whim does not necessarily determine a group. Exercise 4.1 shows the pitfalls of this notion. To reassure you at this point, any presentation of a group in terms of generators and relations in this book will be an honest one; the group in question exists and is uniquely determined. To verify the group properties, especially associativity, is tedious and uninstructive.

_____

4.4   Examples: Groups presented in terms of generators and relations

(a)   The cyclic group $\mathbf{Z}_n$ of order n may be written as

$$\langle g : g^n = 1 \rangle.$$

(b)   The Klein four-group V_4 may be expressed as

$$\langle a, b : a^2 = b^2 = 1, ba = ab \rangle.$$

(c)   The group of symmetries of the regular polygon of n sides is called
the *dihedral group of degree* n, and is given by

$$\langle r, c : r^n = 1, c^2 = 1, cr = r^{-1}c \rangle.$$

The dihedral group in Example 4.4c is denoted by D_n, but note that its
order is $2n$, not n. Of course, the group in Examples 1.2 and 3.8 will now
be denoted D_3. A particularly important example for us will be D_4, the
group of the square, which has already made appearances in Exercises 1.4
and 3.4. Since we'll use it so often, we'll make it explicit here.

4.5   Example: The group of the square

The group D_4 of symmetries of the square in the plane is given by

$$\langle r, c : r^4 = c^2 = 1, cr = r^{-1}c \rangle.$$

Returning to homomorphisms, for $\phi : G \rightarrow G^*$ and $H \subseteq G$, the *image of
H under ϕ* is the subset $\{\phi(g) : g \in H\}$ of G^*. Similarly, recall that the
kernel of a linear transformation is the set of vectors whose image is the
zero vector. A parallel concept for homomorphisms is as follows.

4.6   Definition: Kernel of a homomorphism

If $\phi : G \rightarrow G^*$ is a homomorphism, then the *kernel of ϕ*, denoted $\ker(\phi)$,
is the set $\{g \in G : \phi(g) = e^*\}$, where e^* is the identity of G^*.

Now recall that the kernel of a linear transformation is a vector subspace;
likewise, the kernel of a homomorphism is a subgroup of G. This is easily
seen, in view of Proposition 2.8, by our observing that if $x, y \in \ker(\phi)$, then
$\phi(xy) = \phi(x)\phi(y) = e^*e^* = e^*$ and $\phi(x^{-1}) = (\phi(x))^{-1} = e^{*-1} = e^*$, so xy
and x^{-1} are in the kernel of ϕ. Note that Proposition 4.2 tells us that the
kernel is nonempty.

But the kernel of a homomorphism has a more important property. Suppose that $x \in \ker(\phi)$ and that g is an arbitrary element of the group G on which ϕ has been defined. Then

$$\phi(g^{-1}xg) = \phi(g)^{-1}\phi(x)\phi(g)$$

because ϕ preserves products; but $\phi(x) = e^*$, so

$$\phi(g^{-1}xg) = \phi(g)^{-1}\phi(g) = e^*,$$

and $g^{-1}xg \in \ker(\phi)$ also. Why does this matter? Consider the following.

4.7 Definition: Conjugation of elements in a group

Let $g \in G$. Then *conjugation* by g of an element $x \in G$ is the function carrying x to $g^{-1}xg$. The element $g^{-1}xg$ is a *conjugate* of x.

4.8 Proposition: Conjugation is an action

Conjugation is an action of a group on the set of its own elements.

Proof: Let $g \in G$ (considered as a group) and $x \in G$ (considered as a point set Ω). Certainly conjugation by g sets up a correspondence from x to an element $g^{-1}xg \in G$, so we need only to verify that the two conditions in Definition 3.1 are satisfied. Write x^g for $g^{-1}xg$ to conform to the notation of Definition 3.1. For $g, h \in G$,

$$(x^g)^h = h^{-1}(x^g)h = h^{-1}(g^{-1}xg)h = (gh)^{-1}x(gh) = x^{gh},$$

using Proposition 2.5 and Property 2.1b, and if e is the identity of G, then

$$x^e = e^{-1}xe = exe = e,$$

as required. $\qquad ab$

$$bc$$

$\square$

4.9 Proposition: Conjugation determines an equivalence relation

Conjugation determines an equivalence relation on the set G.

Proof: Let $x, y, z \in G$. Then x is conjugate to itself since $x = e^{-1}xe = exe$; hence, conjugation is reflexive. If x is conjugate to y, then we have some $g \in G$ such that $g^{-1}xg = y$. But then $(g^{-1})^{-1}y(g^{-1}) = x$, so y is conjugate to x as well, and conjugation is symmetric. If x is conjugate to y and y to z, then we have $g, h \in G$ such that $g^{-1}xg = y$ and $h^{-1}yh = z$; but then

$$(gh)^{-1}x(gh) = h^{-1}(g^{-1}xg)h = h^{-1}yh = z,$$

so x is conjugate to z, and conjugation is transitive. This completes the proof. $\square$

4.10 Definition: Conjugate class and centralizer

Consider conjugation both as an action of the group G on the set G and also as an equivalence relation on G. The equivalence class of an element x is the orbit of x under G; and is called the *conjugate class of x*. The stabilizer of an element x under the action of conjugation is called the *centralizer* of x under G.

Next we extend the idea of conjugation from an element to a subgroup of G.

4.11 Proposition: The conjugate of a subgroup is a subgroup

If $H \leq G$ and if $g \in G$, then $g^{-1}Hg$ is a subgroup of G.

Proof: First, we need to make explicit the notation:

$$g^{-1}Hg = \{g^{-1}hg : h \in H\}.$$

(See Exercise 4.2 for an illustration.) Then the details are left for an exercise. □

That phrase "left as an exercise" is one that chills students or readers, but it is not really a bad idea, if used with prudence. The earliest occurrence of such a phrase that the author knows of is in *De Triangulis Omnimodis* by Regiomontanus (written 1464, published 1533) where, following Proposition 22, he says

> This is seen to be the converse of the preceding. Moreover, it has a straightforward proof, as did the preceding. Whereupon I leave it to you for homework.

Despite the potential for chills, we do leave the details of the proof of Proposition 4.11 to you as homework!

Note that, by closure, if $g \in H$, then $g^{-1}Hg \subseteq H$. That the equality also holds under the condition "for all $g \in G$" will be shown in Lemma 5.8.

The fact that a conjugate of a subgroup is again a subgroup and that, as we shall show in the next section, it has the same finite order is important. But a further question is: When is the conjugate of a subgroup actually equal to the subgroup with which we started? Because it is an important question, we'll make the concept explicit.

4.12 Definition: Normal subgroup

Let $H \leq G$; then H is called a *normal subgroup* of G if

$$g^{-1}Hg = H$$

for every $g \in G$. We write $H \trianglelefteq G$ to denote that H is a normal subgroup of G.

The older word for normal subgroups was *invariant*; this was a happier term in its distinguishing what was important (invariance under conjugation) rather than rendering an implied judgment in the term *normal*, but English mathematical usage has accepted *normal* as the standard term.

The discussion following the definition of the kernel in Section 4.6 has established the following proposition.

osition: The kernel is a normal subgroup

a homomorphism $\phi : G \to G^*$ is a normal subgroup of G.

By closure, G is always a normal subgroup of itself; moreover, the subgroup $\{1\}$ consisting of the identity alone is always a normal subgroup. These two normal subgroups always exist (and, of course, coincide in the special case of $|G| = 1$). Some groups have no other normal subgroups than these.

4.14   Definition: Simple group

A group G is called *simple* if its only normal subgroups are G itself and $\{1\}$.

In view of Theorem 3.10 and because the only divisors of a prime p and 1 are p itself, we have the following proposition.

4.15   Proposition: $\mathbf{Z}_p$ is simple

If p is a prime, then the cyclic group $\mathbf{Z}_p$ is simple.

Another important example of a normal subgroup of a group is the set of elements that commutes with all elements of the group. In some instances this set may coincide either with G or $\{1\}$; in others it will be a different normal subgroup.

4.16   Definition: Center of a group

The *center* of a group G is denoted by $C(G)$ and defined by

$$C(G) = \{c \in G : cg = gc \text{ for every } g \in G\}.$$

To verify that $C(G) \trianglelefteq G$, you must check both that $C(G)$ is a subgroup and also that Definition 4.12 is satisfied.

We'll conclude this section with a characterization of normal subgroups and a useful consequence. For the first result, compare Exercises 3.8 and 3.9.

4.17   Proposition: A characterization of normal subgroups

A subgroup H of a group G is normal if and only if $Hg = gH$ for every $g \in G$.

4.18   Proposition: A subgroup of index 2 is normal

If $H \leq G$ and $[G : H] = 2$, then $H \trianglelefteq G$.

Proof: If $[G : H] = 2$ and $g \notin H$, then $G = H \cup Hg$ and $G = H \cup gH$; hence, Proposition 4.17 applies.   □

Exercises

4.1.   Suppose that G is said to be given by

$$\langle x, y : x^3 = y^3 = 1, yx = x^2y \rangle.$$

A group of order 9 would presumably result, but show that in fact these relations are contradictory. (Remember that in such a presentation, $x^3 = 1$ implies that $x \neq 1$ and $x^2 \neq 1$.)

4.2.   In the group D_4 of Example 4.5, let $H = \{1, c\}$. Find the subgroups $r^{-1}Hr$, $r^{-2}Hr^2$, and $(rc)^{-1}H(rc)$.

4.3.   Carry out the details of Proposition 4.11.

4.4.   Show that, if G is an abelian group, then every subgroup of G is normal.

4.5. The *quaternion group* Q_2 may be given by

$$\langle x, y : x^4 = 1, y^2 = x^2, yx = x^{-1}y \rangle.$$

It has order 8. Find all of its subgroups and show that all are normal, but Q_2 is nonabelian. Thus, the converse of Exercise 4.4 is false.

4.6. Let p be a prime; show that the cyclic group $\mathbf{Z}_p$ of order p is simple but that the cyclic group $\mathbf{Z}_{p^2}$ of order p^2 is not simple.

4.7. Find the centers of the groups D_3 and D_4.

4.8. Let $\phi : G \to G^*$ be a homomorphism; show that the image $\phi(G)$ is a subgroup of G^*.

4.9. If $g \in C(G)$, what can you say about the row and column corresponding to g in a group table?

4.10. Find the center of the group Q_2 introduced in Exercise 4.5.

4.11. Prove Proposition 4.17.

4.12. If G is an abelian group, what can you say about $C(G)$?

Chapter 5

Isomorphisms and Automorphisms

In Chapter 1 we noted that Examples 1.1 and 1.2 were genuinely the same in that they each had six elements and that the product tables were identical except for the symbols used to represent the elements. In this chapter we'll make this idea more precise and will extend the discussion.

To start from the opposite end of the spectrum, note that if G and G^* are arbitrary groups and if e^* is the identity of G^*, then the function $\phi : G \to G^*$ defined by $\phi(g) = e^*$ for all $g \in G$ is a homomorphism. Of course, since the image is $\{e^*\}$, all of the structure of G has been suppressed, but the definition of homomorphism has indeed been satisfied and, in fact, we cannot entirely dismiss this apparent pathology.

An intermediate example would come from considering $G = G^* = V_4$ (see Example 4.4b) and taking

$$\phi(a) = \phi(ab) = a \quad \text{and} \quad \phi(1) = \phi(b) = 1.$$

You can easily check that ϕ is a homomorphism and that ϕ carries V_4 to a proper subgroup of itself that is, still, not just $\{1\}$. This leads us to the following definition.

5.1 Definition: Isomorphism

A homomorphism $\phi : G \to G^*$ is an *isomorphism* if ϕ is a one-to-one function.

We may abbreviate *one-to-one* as $1:1$.

If G is a finite group and $\phi : G \to G^*$ is an isomorphism, then clearly $|G| = |\phi(G)|$, but since we did not require ϕ to be a function *onto* G^*, it is possible that the image $\phi(G)$ is a proper subgroup of G^*. However, in this latter instance, some of G^* would be left unaccounted for; hence, when we want to make the sort of comparison we made between Examples 1.1 and 1.2, we want to be sure that all of G^* is used up. Therefore, we say the following.

5.2 Definition: Isomorphic groups

Groups G and G^* are *isomorphic* if there is an isomorphism ϕ from G *onto* G^*. To express the fact that G is isomorphic to G^*, we write $G \cong G^*$.

Now a group consists of a set of elements together with an operation satisfying certain conditions (namely, those of Definition 2.1), and an isomorphism preserves both the set of elements (possibly with a change of symbolism) and the operation (including identity and inverses). Thus, we have made rigorous the notion from Chapter 1 that the meaning of isomorphic groups is that they are exactly the same except for the symbols used to express them. This is made explicit in the idea of an *abstract group*.

5.3 Definition: Abstract group

If G and G^* are isomorphic, we say that they represent the same *abstract group*; thus, two distinct abstract groups are nonisomorphic.

The comparison of Example 1.2, which we now know as the group D_3, with Example 1.3, the group $\mathbf{Z}_6$, shows that these two represent different groups. The question posed at the end of Chapter 1 is, then, whether or not these two are the *only* abstract groups of order 6.

A special case of isomorphism is that of a group onto itself.

5.4 Definition: Automorphism

An isomorphism of a group G onto itself is called an *automorphism* of G.

Certainly the identity function

$$\iota(g) = g \qquad \text{for all } g \in G$$

is an automorphism, but in general we expect to find others.

5.5   Examples: Automorphisms

(a)   The infinite cyclic group $\mathbf{Z}_\infty = \langle x \rangle$ has two automorphisms, the identity and the function determined by $\phi(x) = x^{-1}$ (recall Example 4.3).

(b)   The group $\mathbf{Z}_4 = \langle y \rangle$ also has two automorphisms, the identity and the function determined by $\psi(y) = y^3$.

(c)   The group $\mathbf{Z}_5 = \langle z \rangle$ has four automorphisms, which we may denote by $\phi_k(z) = z^k$ for $k = 1, 2, 3, 4$. Here ϕ_1 is the identity function.

Now consider the composition of two automorphisms: if ϕ and ψ are automorphisms of a group G, then we take

$$(\phi\psi)(g) = \phi(\psi(g)) \qquad \text{for each } g \in G.$$

It is routine to verify that $\phi\psi$ is again an automorphism of G and that, in fact, Definition 2.1 is satisfied when we take the identity function ι for part c of that definition, and, in part d, consider ϕ^{-1} defined by

$$\phi^{-1}(x) = y \qquad \text{if and only if} \qquad \phi(y) = x.$$

Thus, we have the following.

5.6   Proposition: Automorphism group

The automorphisms of a group G form a group under the operation of composition of functions; it is denoted by $\mathbf{A}(G)$.

Connecting the idea of an automorphism group with that of an abstract group, we observe in Examples 5.5a–c,

$$\mathbf{A}(\mathbf{Z}_\infty) \cong \mathbf{Z}_2;$$
$$\mathbf{A}(\mathbf{Z}_4) \cong \mathbf{Z}_2;$$
$$\text{and} \qquad \mathbf{A}(\mathbf{Z}_5) \cong \mathbf{Z}_4.$$

To see that $\mathbf{A}(\mathbf{Z}_5)$ is cyclic, as claimed, and not isomorphic to V_4, note that $\phi_2(\phi_2(z)) = \phi_2(z^2) = z^4 = \phi_4(z)$, that $\phi_2(\phi_4(z)) = \phi_2(z^4) = z^8 = z^3 = \phi_3(z)$, and that $\phi_2(\phi_3(z)) = \phi_2(z^3) = z^6 = z = \phi_1(z)$, so that

$$\phi_2^2 = \phi_4, \qquad \phi_2^3 = \phi_3, \qquad \text{and} \qquad \phi_2^4 = \phi_1 = \iota,$$

where ι, as before, denotes the identity automorphism.

An especially important example of an automorphism of a group G is conjugation, which we discussed in Chapter 4. That discussion now resumes.

5.7 Proposition: Inner automorphisms form a normal subgroup

The conjugation of a group by an element g of G defines an automorphism of G, which we denote by σ_g and call an *inner automorphism* of G. The set of all inner automorphisms of G forms a normal subgroup $\mathbf{I}(G)$ of $\mathbf{A}(G)$, where $\sigma_g^{-1} = \sigma_{g^{-1}}$.

Proof: To show that σ_g, defined by $\sigma_g(x) = g^{-1}xg$ for each $x \in G$, is an automorphism, we must show that σ_g is $1 : 1$, is onto, and preserves products. To show that $\mathbf{I}(G)$ is a subgroup of $\mathbf{A}(G)$, we must show that the composition $\sigma_g \sigma_h$ (where $g, h \in G$) is again an inner automorphism of G and that σ_g has an inverse that is again an inner automorphism. To verify that $\mathbf{I}(G)$ is normal in $\mathbf{A}(G)$, we must show that if τ is *any* automorphism of G, then $\tau^{-1}\sigma_g\tau$ is again an inner automorphism of G. The details are left as Exercise 5.4. □

The preceding sketch of a proof actually asked you to show that if $\sigma_g \in \mathbf{I}(G)$ and if $\tau \in \mathbf{A}(G)$, then $\tau^{-1}\sigma_g\tau$ is again in $\mathbf{I}(G)$. Doing so will, in fact, show that the subgroup $\tau^{-1}\mathbf{I}(G)\tau \subseteq \mathbf{I}(G)$. Since equality of sets means containment in both directions, you also need to know that $\tau^{-1}\mathbf{I}(G)\tau \supseteq \mathbf{I}(G)$. This reverse containment actually follows from the fact that conjugation is an automorphism of G, but it is useful to observe that, in this instance, one containment implies the other.

5.8 Lemma: $g^{-1}Hg \subseteq H$ for every g implies $H \trianglelefteq G$

If H is a subgroup of G having the property that $g^{-1}Hg \subseteq H$ for every $g \in G$, then $H \trianglelefteq G$.

Proof: Suppose that $y \in G$ and that $g^{-1}Hg \subseteq H$ for every $g \in G$. We must show that $H \subseteq y^{-1}Hy$; since y is chosen arbitrarily, we'll know then that $H \subseteq g^{-1}Hg$ for every g. Hence, that equality holds, as required for Definition 4.12. Now $g^{-1}Hg \subseteq H$ holds in particular when $g = y^{-1}$, so

$$yHy^{-1} = (y^{-1})^{-1}Hy^{-1} \subseteq H.$$

Hence, $H = eHe = (y^{-1}y)H(y^{-1}y) = y^{-1}(yHy^{-1})y \subseteq y^{-1}Hy$, as was to be shown. $\square$

Now we'll observe that, since conjugation is an automorphism on G, it is an isomorphism on the set of subgroups of G. In view of Proposition 4.11, we have the following result.

5.9 Proposition: Conjugate subgroups are isomorphic

If $H \leq G$ and $g \in G$, then $g^{-1}Hg \cong H$.

An interesting case not only of isomorphism but even of equality arises with regard to stabilizers when a group acts on a point set. An example of the following proposition is given in Exercise 5.9.

5.10 Proposition: $\alpha^g = \beta$ implies $g^{-1}G_\alpha g = G_\beta$

Let G act on Ω, let $\alpha, \beta \in \Omega$, and let $g \in G$. Then,

$$\alpha^g = \beta \text{ implies that } g^{-1}G_\alpha g = G_\beta.$$

Proof: Let $h \in G$. Then, $h \in G_\beta$

$$\begin{aligned} &\text{iff} && \beta^h = \beta \\ &\text{iff} && \alpha^{gh} = \alpha^g \end{aligned}$$

iff $\alpha^{ghg^{-1}} = \alpha$

iff $ghg^{-1} \in G_\alpha$

iff $h \in g^{-1}G_\alpha g$.

This chain of logical equivalences establishes the equality.   □

5.11   Example: The quaternion group Q_2

In Exercise 4.5 we presented Q_2 as $\langle x, y : x^4 = 1, y^2 = x^2, yx = x^{-1}y \rangle$. Historically, this group arises from the *quaternions*, introduced by Sir William Rowan Hamilton (1788–1856); in this context,

$$Q_2 = \{1, -1, i, -i, j, -j, k, -k\}$$

with the following table:

	1	-1	i	-i	j	-j	k	-k
1	1	-1	i	-i	j	-j	k	-k
-1	-1	1	-i	i	-j	j	-k	k
i	i	-i	-1	1	k	-k	-j	j
-i	-i	i	1	-1	-k	k	j	-j
j	j	-j	-k	k	-1	1	i	-i
-j	-j	j	k	-k	1	-1	-i	i
k	k	-k	j	-j	-i	i	-1	1
-k	-k	k	-j	j	i	-i	1	-1

If we set $\phi(x) = i$ and $\phi(y) = j$ and extend as in Example 4.3, we can check that ϕ is an isomorphism from Q_2 onto itself.

Exercises

5.1.   Let $i = \sqrt{-1}$ (as usual) and let $G = \langle A, B \rangle$ where

$$A = \begin{bmatrix} i & 0 \\ 0 & i \end{bmatrix}, \qquad B = \begin{bmatrix} 0 & 1 \\ 1 & 0 \end{bmatrix}.$$

Express G as an abstract group by giving a presentation of G in terms of generators and relations.

5.2.   Let $H = \langle C, D \rangle$, where

$$C = \begin{bmatrix} 0 & 1 \\ -1 & 0 \end{bmatrix}, \qquad \text{and} \qquad D = \begin{bmatrix} 1 & 0 \\ 0 & -1 \end{bmatrix}.$$

Find the abstract group to which H is isomorphic.

5.3.   Using the composition of functions as in Proposition 5.6, show that isomorphism between groups is an equivalence relation.

5.4.   ~~Fill in the details of the~~ *prove* ~~proof of~~ Proposition 5.7.

5.5.   If $\mathbf{Z}_\infty = \langle x \rangle$, find an isomorphism τ from $\langle x \rangle$ onto $\langle x^2 \rangle$. Show that τ is *not* an automorphism of $\mathbf{Z}_\infty$, even though $\langle x \rangle \cong \langle x^2 \rangle$.

5.6.   If G is abelian and $H \leq G$, what can you say about the conjugates of H?

5.7.   Show that the converse of Proposition 5.9 is false by finding

   a.   Subgroups of V_4 that are isomorphic but not conjugate;

   b.   Subgroups of D_4 that are isomorphic but not conjugate.

(*Hint*: Consider Exercises 5.6, 4.4, and 4.7.)

5.8.   If G is an abelian group, what can you say about $\mathbf{I}(G)$?

5.9.   Illustrate Proposition 5.10 for the group D_4 of the square bipyramid (Figure 5.1) by taking α and β to be vertices 2 and 5, respectively. Specifically, find G_2 and G_5, find the two elements which carry 2 to 5, and show that conjugation by each of these elements transforms G_2 into G_5.

5.10. Using the composition of functions in Proposition 5.6, show that isomorphism between groups is an equivalence relation.

5.11. Verify that the function ϕ in Example 5.11 is an isomorphism onto.

5.12. Let $\mathbf{Z}_\infty = \langle x \rangle$; verify that $\mathbf{A}(\mathbf{Z}_\infty) \cong \mathbf{Z}_2$. (See Exercise 5.5.)

5.13. Find all automorphisms of the group Z_3. What abstract group is $\mathbf{A}(Z_3)$?

5.14. Find all automorphisms of the group D_3. What abstract group is $\mathbf{A}(D_3)$?

5.15. Find all automorphisms of the group V_4. What abstract group is $\mathbf{A}(V_4)$?

5.16. Show that $\mathbf{A}(D_4) \cong D_4$.

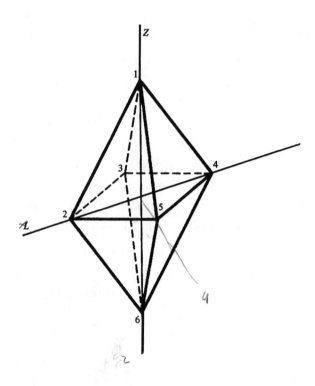

FIGURE 5.1
Square bipyramid

Chapter 6

Factor Groups

In Chapter 3 we introduced the idea of cosets and considered some of their properties, including the important criterion in Proposition 3.7 for equality of right cosets:

If $H \leq G$ and $x, y \in G$, then $Hx = Hy$ if and only if $xy^{-1} \in H$.

In this chapter we'll treat the distinct (right) cosets in a group G determined by a given subgroup H of G as the elements of a new group. First, we'll need to consider the related concept of a formal product of subgroups.

6.1 Definition: Formal product of subgroups

Let $H, K \leq G$; we define the *formal product* of H and K by

$$HK = \{hk : h \in H, k \in K\}.$$

By closure, HK is a subset of G, but it will not necessarily be a subgroup; an example is given in Exercise 6.1. However, here is the definitive criterion for HK to be a subgroup of G.

6.2 Proposition: $HK \leq G$ if and only if $HK = KH$

Let $H, K \leq G$; then $HK \leq G$ if and only if $HK = KH$.

Proof: First we assume that $HK = KH$. We'll use Proposition 2.8 to show that HK is a subgroup of G. Since H and K are subgroups of G, the identity e of G is in both, and by Definition 6.1, $e = ee \in HK$. The crucial point now is that the equality $HK = KH$ means that any element kh of KH can also be written as an element of HK; that is, there must exist elements $h^* \in H$ and $k^* \in K$ such that $kh = h^*k^*$. (An illustration is given in Exercise 6.2.) Now let $h_1, h_2 \in H$ and $k_1, k_2 \in K$. The equality $HK = KH$ then ensures that there exist $h_2^* \in H$ and $k_1^* \in K$ such that $k_1 h_2 = h_2^* k_1^*$; hence,

$$(h_1 k_1)(h_2 k_2) = h_1 (k_1 h_2) k_2 = h_1 (h_2^* k_1^*) k_2 = (h_1 h_2^*)(k_1^* k_2) \in HK$$

by closure in H and K, and thus Proposition 2.8b is satisfied. Now for $hk \in HK$, we need to show that $(hk)^{-1} \in HK$. But again, the equality $HK = KH$ ensures that

$$(hk)^{-1} = k^{-1}h^{-1} = h^*k^* \in HK$$

for some $h^* \in H$ and $k^* \in K$. Thus, Proposition 2.8c is satisfied also, and HK is a subgroup of G. For the converse, we assume that HK is a subgroup of G; we must prove that $HK = KH$. Now since HK is a subgroup of G, each element of HK is the inverse of a unique element of HK (recall Exercise 2.16); hence, for HK a complete list of inverses is also a complete list of elements, that is,

$$\begin{aligned} HK &= \{(hk)^{-1} : h \in H, k \in K\} \\ &= \{k^{-1}h^{-1} : h \in H, k \in K\} \\ &= KH \end{aligned}$$

since each element of H is an inverse of some element of H, and similarly for K. This completes the proof.   □

An important special case of Proposition 6.2 follows; the proof is left as Exercise 6.3.

6.3   Proposition: $HK \leq G$ when H is normal in G

Let $H \trianglelefteq G$ and $K \leq G$. Then $HK \leq G$.

Continuing with the development, let H be a subgroup of G; by Theorem 3.6 we know that we have a decomposition of G into distinct cosets. In particular, if G is finite, we can write

$$G = Hg_1 \cup Hg_2 \cup \cdots \cup Hg_n$$

where $n = [G : H]$ by Proposition 3.9 and Theorem 3.12. Regardless of whether or not G is finite, we can form the *factor set* as follows.

6.4 Definition: Factor set G/H

If $H \leq G$, then the *factor set* G/H (read G mod H) is defined to be the set of distinct right cosets of H in G.

Now our object is to treat the distinct cosets as the elements of a group formed from the factor set; to do so, we need to define a group operation. The most natural way in which to specify a product on the factor set G/H would be to set $(Hx)(Hy) = H(xy)$ for $x, y \in G$. Since the associative property holds in G, we could simply write Hxy for this product. However, in light of the fact that Hx and Hy may be the same coset (according to Proposition 3.7, this occurs precisely when $xy^{-1} \in H$), we need to make sure that this product is *well-defined*, that is,

if $Hx = Hy$ and $Hu = Hv$, then $Hxu = Hyv$.

To illustrate this concept in a more elementary context, consider for a moment the rational numbers $\mathbf{Q}$. If we were to define an operation $\#$ on $\mathbf{Q}$ by $a/b \# c/d = (a + c)/(b + d)$, the result would be that $1/2 \# 1/3 = 2/5$ whereas $2/4 \# 1/3 = 3/7$, even though $1/2 = 2/4$; thus $\#$ would not make sense for $\mathbf{Q}$ — the operation $\#$ would not be well-defined. See Exercise 6.4 for a further illustration.

Now let's assume that $H \trianglelefteq G$; we'll show that normality is sufficient to make the operation in G/H well-defined. Let $Hx = Hy$ and $Hu = Hv$, where $x, y, u, v \in G$. By Proposition 3.7 we know that xy^{-1}, $uv^{-1} \in H$. Since $H \trianglelefteq G$, we know also that $y(uv^{-1})y^{-1} \in H$. Using Proposition 3.5 at the second and fourth lines below, we now have

$$\begin{aligned}
(Hy)(Hv) &= Hyv \\
&= H(yuv^{-1}y^{-1})yv \\
&= Hyuv^{-1}v = Hyu \\
&= H(xy^{-1})yu \\
&= Hxu = (Hx)(Hu),
\end{aligned}$$

which completes the check that the proposed operation is well-defined. In fact, we have now proved the following.

6.5 Proposition: If $H \trianglelefteq G$, then G/H is a group

Let H be a normal subgroup of G. Then the factor set G/H forms a group under the operation $(Hx)(Hy) = Hxy$.

Note that with the specified operation in the factor group, we have $(H)(Hg) = Hg = (Hg)(H)$ (since $H1 = H$); thus H itself is the identity of G/H. This fact is so important that it will be repeated here as a proposition.

6.6 Proposition: If $H \trianglelefteq G$, then H is the identity of G/H

Let H be a normal subgroup of G. Then H itself is the identity element of G/H; that is, $(Hg)(H) = (H)(Hg) = H$ for every $g \in G$.

An important function relates a group G to such a factor group G/H.

6.7 Definition: The canonical homomorphism from G to G/H

Let $H \trianglelefteq G$, and define $\eta : G \to G/H$ by $\eta(g) = Hg$. Then η is called the *canonical* or *natural homomorphism* from G to G/H. (The fact that η is a homomorphism has been left as Exercise 6.5.)

Now we are ready to see how factor groups figure in three classic theorems about homomorphisms.

6.8 Theorem: ~~Homomorphism~~ *Isomorphism* theorem I

Let $\phi : G \to G^*$ be a homomorphism of G *onto* G^* and let K be the kernel of ϕ. Then $K \trianglelefteq G$, and G/K forms a group under the operation $(Kg)(Kh) = Kgh$. Moreover, $G/K \cong G^*$.

Proof: In Proposition 4.13 we showed that K is a normal subgroup of G. From Proposition 6.5 we know that G/K forms a group under the operation specified. It remains to exhibit an isomorphism from G/K onto G^*. Define $\sigma : G/K \to G^*$ by $\sigma(Kg) = \phi(g)$ for each coset Kg. Then

$$\begin{aligned}
\sigma((Kg)(Kh)) &= \sigma(Kgh) \\
&= \phi(gh) \\
&= \phi(g)\phi(h) \\
&= \sigma(Kg)\sigma(Kh),
\end{aligned}$$

so σ is a homomorphism from G/K into G^*. Now σ is onto because $\phi(G) = G^*$. That σ is both well-defined and one-to-one is shown from the following, where e^* denotes the identity of G^*.

$$\begin{aligned}
\sigma(Kg) = \sigma(Kh) \quad & \text{iff } \phi(g) = \phi(h) \\
& \text{iff } \phi(g)\phi(h)^{-1} = e^* \\
& \text{iff } \phi(gh^{-1}) = e^* \\
& \text{iff } gh^{-1} \in K \\
& \text{iff } Kg = Kh.
\end{aligned}$$

This completes the proof of Theorem 6.8.   □

6.9 Example: A factor group constructed from D_4

Write the group D_4 in the notation of Example 4.5 and the group V_4 in the notation of Example 4.4b. Let $\phi(r) = a$ and $\phi(c) = b$; as in Example 4.3, we extend this definition to powers and products of r and c. You should check that ϕ preserves the three relations in Example 4.5 and that the kernel of ϕ is $K = \{1, r^2\}$. We then have

$$
\begin{aligned}
Kr &= \{r, r^3\} = Kr^3, \\
Kc &= \{c, r^2c\} = Kr^2c, \\
Krc &= \{rc, r^3c\} = Kr^3c, \\
K &= Kr^2,
\end{aligned}
$$

and writing $D_4/K = \{K, Kr, Kc, Krc\}$, we can easily see the isomorphism with V_4.

6.10   Theorem: Homomorphism theorem II

Let $\phi : G \to G^*$ be a homomorphism of G *onto* G^* with kernel K. Then for each subgroup H of G containing K, we have $H/K \cong \phi(H)$. Conversely, if H^* is a subgroup of G^*, then there is a subgroup H of G containing K for which $\phi(H) = H^*$ and $H/K \cong H^*$. Moreover, if $K \leq H \leq G$, then $H \triangleleft G$ if and only if $\phi(H) \triangleleft G^*$.

Proof: First, let $K \leq H \leq G$. By closure in the subgroup H, a coset $Kx \subseteq H$ if and only if $x \in H$. Hence, H is a union of cosets of the form Kx with $x \in H$. Consider ϕ as a function with its domain restricted (limited) to H; then ϕ is a homomorphism from H onto $\phi(H)$ with kernel K, and by Theorem 6.8, $H/K \cong \phi(H)$. Now let $H^* \leq G^*$, and set $H = \{g \in G : \phi(g) \in H^*\}$; H is then called the *inverse image* of H^* under ϕ. Then if $k \in K$, we have $\phi(k) = e^*$, the identity of G^*, and $e^* \in H^*$; hence, $K \leq H$. That $\phi(H) = H^*$ is immediate from the definition of H. As before, if the domain of ϕ is restricted to H, then Theorem 6.8 shows that $H/K \cong H^*$. To consider the question of normality, let $K \leq H \triangleleft G$, and let $H^* = \phi(H)$. If $x \in H^*$, then $x = \phi(y)$ for some $y \in H$, and if $u \in G^*$, then $u = \phi(v)$ for some $v \in G$ (since ϕ carries G onto G^*). Now

$$
u^{-1}xu = \phi(v)^{-1}\phi(y)\phi(v) = \phi(v^{-1}yv) \in \phi(H) = H^*
$$

since $v^{-1}yv \in H$ by normality of H in G. Hence (recall Lemma 5.8) H^* is normal in G^*. The proof that $H^* \triangleleft G^*$ implies that $H \triangleleft G$ is left as Exercise 6.6.   □

This theorem says, in effect, that the pattern of subgroups of G^* is precisely the same as the pattern of subgroups of G containing the kernel K of ϕ. In terms of the preceding example, we have the pattern of inclusions shown in Figure 6.1; the solid lines show the pattern of inclusions

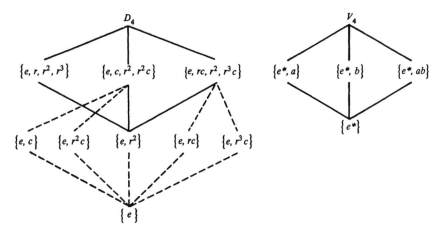

FIGURE 6.1
Subgroups of D_4

that are carried over by ϕ, and the dotted lines indicate inclusions that are
suppressed because they do not pertain to subgroups containing the kernel
$\{e, r^2\}$.

6.11   Theorem: Homomorphism theorem III

Let H be a subgroup of a group G and N be a normal subgroup of G.
Then,

NH is a subgroup of G,
N is a normal subgroup of NH,
$H \cap N$ is a normal subgroup of H, and
$NH/N \cong H/(H \cap N)$.

Proof: First, $NH \leq G$ by Proposition 6.3. Next, N is normal in NH
because N is normal in all of G, of which NH is a subgroup. To see
that $H \cap N$ is normal in H (although not necessarily normal in G itself),
let $x \in H \cap N$ and $y \in H$; we must show that $y^{-1}xy \in H \cap N$. But
$y^{-1}xy \in H$ by closure and $y^{-1}xy \in N$ because $N \trianglelefteq G$. Finally, define
$\psi : H \to NH/N$ by $\psi(h) = Nh$ for each $h \in H$. Now h is in the kernel
of ψ if and only if $\psi(h)$ is the identity of NH/N, which by Proposition 6.6
is N itself. But, recalling Proposition 3.5, that means that $h \in N$. We
have established the conditions of Theorem 6.8 with $H \cap N$ as the kernel
of ψ; hence, $NH/N \cong H/(H \cap N)$. This completes the proof of Theorem
6.11. $\square$

This theorem is especially useful in the study of group extensions, in which, given the kernel N and an isomorphic copy of the factor group G/N, one attempts to find the groups G that can have the given kernel and the given factor group.

Exercises

6.1. For D_4 denoted as in Example 4.5, let $H = \langle c \rangle$ and $K = \langle rc \rangle$. Find HK and KH and verify that neither of these subsets is a subgroup of D_4.

6.2. For D_4 denoted as in Example 4.5, let $H = \langle r \rangle$ and $K = \langle c \rangle$. Then, $D_4 = HK = KH$. For each element of D_4 find an expression for that element as a member of HK and an expression for that element as a member of KH.

6.3. Use Proposition 6.2 to prove Proposition 6.3.

6.4. For D_4 denoted as in Example 4.5, let $H = \langle c \rangle$. Find elements g_1, g_2, g_3, g_4 in D_4 for which $Hg_1 = Hg_2$ and $Hg_3 = Hg_4$, but $Hg_1g_3 \neq Hg_2g_4$.

6.5. Verify that the function η given in Definition 6.7 is indeed a homomorphism.

6.6. Complete the proof of Theorem 6.10.

6.7. Let $G = \langle g : g^9 = 1 \rangle$ be cyclic of order 9 and $G^* = \langle h : h^3 = 1 \rangle$ cyclic of order 3. Prove that $\phi(g^a) = h^{2a}$ for $0 \leq a \leq 8$ defines a homomorphism of G onto G^* but that ϕ is not an isomorphism.

6.8. Write the quaternion group Q_2 from Exercise 4.5 and Example 5.11 as

$$\langle x, y : x^4 = 1, y^2 = x^2, yx = x^{-1}y \rangle.$$

Take $K = \{1, x^2\}$ and draw a diagram like Figure 6.1 for Q_2. For the canonical homomorphism $\eta : Q_2 \to Q_2/K$, you have $\eta(Q_2) = Q_2/K$; note that this is *equality* rather than *isomorphism*. Now verify that $\eta(Q_2) \cong V_4$.

6.9.

a. Let $A, B \leq G$ and $g \in G$. Prove that $Ag \cap Bg = (A \cap B)g$.

b. Let $A, B \trianglelefteq G$. If G/A and G/B are both abelian, prove that $G/(A \cap B)$ is abelian.

Chapter 7

Sylow Subgroups

The serious study of finite groups began in 1872 with the publication of some remarkable results by Ludvig Sylow (1822–1918). In this chapter we'll consider these theorems, some related results, and some applications. To concentrate on how the Sylow theorems are *used*, we omit the proofs from this chapter and refer the interested reader to Chapter 22.

Note that the hypotheses of the Sylow theorems will require that G be a finite group. The discussion of normalizers, however, applies to arbitrary groups.

The theorem of Lagrange (Theorem 3.10) says that if H is a subgroup of a finite group G, then $|H|$ divides $|G|$. The logical converse of this result is, if k is a divisor of $|G|$, then G has a subgroup of order k; unfortunately, this statement is false, and in Chapter 8 we'll find a group A_5 of order 60 that has no subgroup of order 15, 20, or 30. The first Sylow theorem is a partial converse to the theorem of Lagrange, although it is not the strongest known result in this direction.

7.1 Theorem: Sylow theorem I

Let p be a prime, $|G| = p^e q$, where p does not divide q. Then G contains a subgroup of order p^e, which is called a *Sylow p-subgroup* of G.

It is especially important to note that the order of a Sylow p-subgroup is p^e, not p itself (unless $e = 1$).

We should make it clear that q may be a prime, a power of a prime, or a composite number. Note also that it is possible to have $e = 0$ in Theorem 7.1 in the event that p is a prime that does not divide $|G|$; then, of course, $q = |G|$, and $\{1\}$ qualifies (vacuously) as a Sylow p-subgroup of G. However, in practice, we'll refer to a Sylow p-subgroup only when p is a prime actually dividing $|G|$, so that $e \geq 1$.

Theorem 7.1 says that the group A_5 mentioned above, whose order is $60 = 2^2 \cdot 3 \cdot 5$, must have subgroups of orders 4, 3, and 5. Later in this chapter, we'll see that it also has subgroups of order 2, and in Chapter 8 we'll discover that it has subgroups of orders 6, 10, and 12 as well.

Groups whose orders are powers of a single prime p have many interesting properties; we call such groups *p-groups* (hence, the term Sylow p-subgroup). Since the order of an arbitrary element of a finite group must divide the order of the group, each element of a finite p-group has some power of p as its order. We'll state and prove two properties of finite p-groups now; in particular, we'll see that the converse of the theorem of Lagrange does hold for finite p-groups.

7.2 Proposition: The center of a finite p-group is non-trivial

If p is a prime and if $|G| = p^n$ for some positive integer n, then the center of G has order greater than 1.

Proof: The center of a group G was defined in Definition 4.16 as the set of elements of G that commute with every element of G. Let G act on itself by conjugation, as in Proposition 4.8, and let

$$A_1 = a_1^G, \qquad A_2 = a_2^G, \qquad \ldots, \qquad A_k = a_k^G$$

be the distinct conjugate classes of G, the orbits of the set G under the action of conjugation. (Here each a_j is an element of G.) Now for any i, $|A_i| = 1$ if and only if $a_i \in C(G)$, since for arbitrary $g \in G$,

$$a_i^g = a_i \quad \text{if and only if} \quad g^{-1}a_ig = a_i \quad \text{if and only if} \quad a_ig = ga_i.$$

Hence G is the set-theoretic union of $C(G)$ (which is nonempty since at least $1 \in C(G)$) and the union of those orbits A_i that contain more than one element. Consequently,

$$p^n = |G| = |C(G)| + \sum |A_i|,$$

where the summation is taken over all those A_i having order greater than 1. By Theorems 3.10 and 3.13, $|A_i|$ divides $|G|$ and so is a power of p. Thus p divides every $|A_i|$ that is not equal to 1. Since p divides every other term in the preceding equation, it must also divide $|C(G)|$, and hence $|C(G)| > 1$. $\square$

7.3 Proposition: Subgroups of finite p-groups

If $|G| = p^n$, then G has subgroups of orders $1, p, p^2, \ldots, p^n$.

Proof: This proof is given for completeness but is rather involved; if you decide to omit it, at least for the present, you will still be able to continue with the text. If you do study it, the best way in which to understand it is to work through the text here with a specific example; suggestions for doing so are given in Exercises 7.9 and 7.10.

Obviously, G has order p^n and $\{1\}$ has order 1, so we must prove the result for orders p through p^{n-1}. We let $1 \le m < n$, proceed by induction on m, and consider two cases.

Case 1: G is abelian. Let x be any element distinct from 1 in G, and let p^r be the order of the element x (remember that the order of x must be a divisor of p^n). First, if $r = n$, then G is in fact the cyclic group generated by x, and

$$\left(x^{p^{n-m}}\right)^{p^m} = x^{(p^{n-m})(p^m)} = x^{p^{(n-m)+m}} = x^{p^n} = 1. \qquad (*)$$

Thus $x^{p^{n-m}}$ generates a subgroup of order p^m, which is what we need. Now if $r \ne n$, then $r < n$. If $r = m$, then $\langle x \rangle$ has order p^m and is the required subgroup. If $r \ne m$, we have two subcases to consider.

> *Subcase 1a.* Let $r > m$; then $x^{p^{r-m}}$ generates a subgroup of order p^m (check this with an equation like $(*)$).
>
> *Subcase 1b.* Let $r < m$; then since G is abelian, $\langle x \rangle \trianglelefteq G$, and $G/\langle x \rangle$ is a group of order p^{n-r}. Now $m - r < m$, so by the induction hypothesis, $G/\langle x \rangle$ contains a subgroup $H/\langle x \rangle$ of order p^{m-r}, and H itself is a subgroup of order p^m in G. This completes *Case 1.*

Case 2: G is nonabelian. By Proposition 7.2, $C(G)$ is nontrivial; thus there exists an integer t with $1 \le t < n$ such that $|C(G)| = p^t$. If $m = t$, then $C(G)$ itself is the required subgroup. If $m < t$, then the result follows by Case 1 together with Exercise 2.6 (a subgroup of a subgroup is again a subgroup). If $m > t$, then $G/C(G)$ has order p^{n-t} and, by the induction hypothesis, contains a subgroup $H/\langle x \rangle$ of order p^{m-t}, whence H is a subgroup of G having order p^m. This completes the proof of Proposition 7.3. □

Proposition 7.3 now gives us the following corollary to the first Sylow theorem.

7.4   Corollary: Existence of p-subgroups of a finite group

If p is a prime and if $|G| = p^e q$, where p does not divide q, then G has subgroups of orders $1, p, p^2, \ldots, p^e$.

Proof: This follows immediately from Theorem 7.1 and Proposition 7.3, using Exercise 2.6. □

Since the second Sylow theorem concerns conjugacy of subgroups, it will be helpful to return briefly to the idea of conjugacy as a group action. In Proposition 4.11 we showed that a conjugate of a subgroup of G is again a subgroup of G. Thus, we have the following result parallel to that in Proposition 4.8.

7.5   Remark: Conjugation is an action of G on its subgroups

Conjugation is an action of G on the set of its own subgroups.

Now the orbit of a subgroup H under the action of conjugation is just the set of subgroups of G that are conjugate to H. What is the stabilizer of H? From Definition 3.2 and Proposition 3.3 we know that it is a subgroup of G consisting of all those elements of G that carry H to itself under the action of conjugation. Here an element g is in the stabilizer of H when $g^{-1}Hg = H$ *as a set*; we do not ask that $g^{-1}hg = h$ for each element h of H. Because of the importance of the action of conjugation on the set of subgroups, we give a special name to the stabilizers.

7.6   Definition: Normalizer of a subgroup

The stabilizer of a subgroup H of a group G under the action of conjugation is called the *normalizer of H in G* and is denoted by $N_G H$.

7.7 Proposition: $H \trianglelefteq N_G H$

Let $H \leq G$. Then $H \trianglelefteq N_G H$.

Proof: By Proposition 3.3, $N_G H \leq G$. If $h \in H$, then $h^{-1} H h = H$ by closure, so $h \in N_G H$; thus, $H \leq N_G H$. Moreover, H is normal in $N_G H$ because $n \in N_G H$ means precisely that $n^{-1} H n = H$. $\qquad\square$

We're ready for the second Sylow theorem, which relates the concepts of p-subgroup and conjugacy. Note that if G satisfies the hypotheses of Theorem 7.1 and if H is a subgroup of G with $|H| = p^k$, then $k \leq e$ since p^k must divide p^e by Theorem 3.10.

7.8 Theorem: Sylow theorem II

Let $|G| = p^e q$ with p prime and p not dividing q. Let $H \leq G$ with $|H| = p^k$, and let P be *any* Sylow p-subgroup of G. Then there exists an element $g \in G$ such that $H \leq g^{-1} P g$.

Observe first that since conjugation is an automorphism of G and hence a one-to-one function, $g^{-1} P g$ contains p^e elements and hence is also a Sylow p-subgroup of G. (Note that $g^{-1} P g$ is a subgroup of G by Proposition 4.11.) Thus Theorem 7.8 tells us that if H is a subgroup of G containing p^k elements, then H is a subgroup of *some* Sylow p-subgroup of G. More important, if $k = e$ in the hypothesis of Theorem 7.8, then H and P are two arbitrary Sylow p-subgroups of G. Thus we have proved the following corollary.

7.9 Corollary: Sylow p-subgroups are conjugate

For a given prime p, any two Sylow p-subgroups of G are conjugate.

Now by Remark 7.5, conjugacy induces an action of the group G on the set of subgroups of G, and conjugacy (being an isomorphism) preserves order. Hence for a given prime p, conjugacy determines an action of G on the set of Sylow p-subgroups of G, and Corollary 7.9 guarantees that all the Sylow p-subgroups of G lie in a single orbit with respect to this action. Now apply Theorem 3.13 with α as a Sylow p-subgroup of G; then the orbit α^G is precisely the set of Sylow p-subgroups of G for the particular prime p under consideration. The order of α^G is precisely the number of distinct Sylow p-subgroups of G for this p. The number of distinct Sylow p-subgroups is therefore the index in G of the stabilizer of any one Sylow p-subgroup. But we saw above that such a stabilizer is in fact the normalizer of the subgroup in question. We have proved the following two propositions.

7.10   Proposition: Number of Sylow p-subgroups in a group

The number of distinct Sylow p-subgroups of a group G is equal to the index in G of the normalizer of any one Sylow p-subgroup. Moreover, the normalizers of two Sylow p-subgroups have the same order.

7.11   Proposition: Unique Sylow p-subgroups are normal

A Sylow p-subgroup is normal in G if and only if it is the unique Sylow p-subgroup for that prime p.

It is worth adding that two distinct Sylow p-subgroups have distinct normalizers as well. For if P and Q were Sylow p-subgroups of G having the same normalizer N, then P and Q would both be normal in N, but by Corollary 7.9 they are conjugate to one another, so they must in fact be equal. We refer to the normalizer of a Sylow p-subgroup as a *Sylow normalizer*. In summary, we have proved the following.

7.12   Proposition: Sylow normalizers are distinct

If P_1 and P_2 are distinct Sylow p-subgroups of G, then their normalizers are also distinct.

The third of the Sylow theorems gives a different fact about the number of distinct Sylow p-subgroups.

7.13 Theorem: Sylow theorem III

Let $|G| = p^e q$ with p prime and p not dividing q. Then the number s_p of distinct Sylow p-subgroups of G is congruent to 1 modulo p and is a divisor of q.

That s_p is congruent to 1 modulo p means that p divides $s_p - 1$; this is denoted by $s_p \equiv 1 \pmod{p}$.

7.14 Corollary: A Sylow p-subgroup is normal when $s_p = 1$

A group G has a normal Sylow p-subgroup for the prime p if and only if $s_p = 1$.

Proof: This follows immediately from Proposition 7.11. $\square$

The Sylow theorems give a remarkable amount of information about the structure of a finite group of a given order. For example, we'll show that there is only one abstract group of order 35, but first we need the following result.

7.15 Proposition: Normal subgroups with trivial intersection

Let N and M be normal subgroups of G for which $N \cap M = \{1\}$. Then $nm = mn$ for every $n \in N$ and $m \in M$.

Proof: For $n \in N$ and $m \in M$, we have

$$n^{-1}m^{-1}nm = (n^{-1}m^{-1}n)m \in M$$

by closure in M, where $n^{-1}m^{-1}n \in M$ by normality. Similarly,

$$n^{-1}m^{-1}nm = n^{-1}(m^{-1}nm) \in N,$$

so $n^{-1}m^{-1}nm \in N \cap M = \{1\}$. But then $n^{-1}m^{-1}nm = 1$, and so $nm = mn$. $\square$

Note that Proposition 7.15 says that individual elements of N commute with individual elements of M. By Proposition 6.3, MN is a subgroup of G, but MN need not be abelian since either or both of M and N may be nonabelian.

7.16   Example: Group of order 35 must be cyclic

Let $|G| = 35 = 5 \cdot 7$. By Theorem 7.1, G must have at least one subgroup H of order 5 and at least one subgroup K of order 7. Now $s_5 \equiv 1 \pmod{5}$ and s_5 divides 7. The only divisors of 7 are 1 and 7, and 7 is not congruent to 1 mod 5; hence $s_5 = 1$ and, by Corollary 7.14, H is normal in G. Similarly, K is normal in G. Now $H \cap K$ is a subgroup of both H and K and so has order dividing both 5 and 7; hence, $H \cap K = \{1\}$. By Exercise 3.10, H and K are both cyclic; when we write $H = \langle h \rangle$ and $K = \langle k \rangle$, Proposition 7.15 tells us that $hk = kh$. But then,

$$(hk)^2 = h(kh)k = h(hk)k = h^2 k^2,$$

and similarly, $(hk)^n = h^n k^n$ for any power n. Now if $(hk)^n = 1$, then $h^n = k^{-n}$; this element is in both H and K and so must be in the identity element 1 itself; thus both the order 5 of h and the order 7 of k must divide n. This shows that hk has order 35 and so generates G. Therefore, G is cyclic; there is only one abstract group of order 35.

We'll conclude this section with a particularly nice theorem about Sylow normalizers and subgroups containing them.

7.17   Theorem: Sylow normalizers are self-normalizing

Let P be a Sylow p-subgroup of G, $N = N_G P$, and H any subgroup of G containing N (including N itself). Then H is its own normalizer; that is, $N_G H = H$.

Proof: By Proposition 7.7, we need only to show that $N_G H \subseteq H$. Let $g \in N_G H$. Then $g^{-1} P g$ is a Sylow p-subgroup of $g^{-1} H g = H$; that is, both $g^{-1} P g$ and P are Sylow p-subgroups of H. Hence by Corollary 7.9 applied to H rather than to G, there exists some $h \in H$ such that $h^{-1}(g^{-1} P g)h = P$. But then $(gh)^{-1} P(gh) = P$, so $gh \in N$, and since $N \leq H$, we have

$gh \in H$ also. Then by closure in H, $g = (gh)h^{-1} \in H$. This shows that $N_G H \subseteq H$. □

Exercises

7.1. Apply an analysis like that in Example 7.16 to show that there is only one abstract group of order 15. Why does such an argument fail for a group of order 21? (In fact, there is a noncyclic group of order 21.)

7.2. Show that a group of order 39 cannot be simple.

7.3. Prove Proposition 7.11.

7.4. Let N and M be subgroups of G having relatively prime order, but do not assume that they are normal in G. Show that $N \cap M = \{1\}$, but show by a specific example that $nm = mn$ need *not* hold for every $n \in N$ and $m \in M$.

7.5. Let G be abelian and p be a prime dividing $|G|$; show that G has a unique Sylow p-subgroup.

7.6. Find the Sylow 2- and 3-subgroups of $\mathbf{Z}_{24}$.

7.7. What are the Sylow p-subgroups of a finite p-group?

7.8. Let $|G| = 12$ and assume that G does not have a normal Sylow 3-subgroup. Show that G must have exactly eight elements of order 3 and that, therefore, G must have a normal 2-subgroup. This shows that a group of order 12 cannot be simple.

7.9. Go step-by-step through Case 1 of the Proof of Proposition 7.3, taking the specific values $n = 8$ and $m = 5$. For Subcase 1a, use $r = 7$; for Subcase 1b, use $r = 3$.

7.10. Go step-by-step through Case 2 of the Proof of Proposition 7.3, taking the specific values $n = 8$ and $m = 5$. Consider the possibilities $t = 5$, $t = 6$, and $t = 2$.

7.11. Strictly speaking, Equation (∗) in the Proof of Proposition 7.3 only verifies that the order of $x^{p^{n-m}}$ is a divisor of p^m. Show that, in fact, equality holds.

Chapter 8

Permutation Groups

In Examples 1.2, 1.3, and 4.5 we introduced groups D_3, $\mathbf{Z}_6$, and D_4 which entailed symmetries of the trigonal bipyramid, the hexagon, and the square. More generally, Example 4.4c introduced the group D_n of the regular polygon having n sides. (Note that the group $\mathbf{Z}_6$ of rotations of the hexagon is a proper subgroup of the full group D_6 of symmetries of the regular hexagon.) In the notation of Chapter 1 we viewed these groups as permutations among the vertices of the figure. Comparing Exercises 1.4 and 3.4 with Example 4.5, we have $r = (1234)$ and $c = (14)(23)$, which generate the whole of D_4; here we have expressed the group terms of its action on the set of vertices, although, of course, we could have considered the sides of the square instead. (See Exercises 8.1, 3.12, and 3.13.)

By a *permutation* we mean a one-to-one function on a finite set. In fact, we can think of D_4 merely as an abstract group acting on the set $\{1, 2, 3, 4\}$, and similarly, D_3 is a group acting on the set $\{1, 2, 3\}$. There is, however, a notable distinction between D_4 and D_3. All of the possible permutations of the set $\{1, 2, 3\}$ are included in D_3, namely,

$$i, \ (12), \ (13), \ (23), \ (123), \ (132),$$

with i representing the identity. However, some possible permutations of $\{1, 2, 3, 4\}$ do not appear in the list of elements of D_4 as we worked it out in Exercise 1.4; for example, (12) would represent interchanging the two vertices joined by one side of the square while leaving the opposite two vertices fixed, something we could not do with a rigid square. The example from the regular hexagon treated in Example 1.3 has neither all of the possible motions of the figure (for example, $(26)(35)$ represents a flip about a vertical axis but is not included in the cyclic group of rotations) nor all of the possible permutations of the set $\{1, 2, 3, 4, 5, 6\}$. With these examples in mind, we formulate the following.

8.1   Proposition: Symmetric group S_n

The collection of all permutations of the set $\{1, 2, \ldots, n\}$ under the operation of composition of functions forms a group S_n called the *symmetric group of degree n*. This group has order $n!$ (n-factorial).

Proof: The group properties are easy to check. For the order, consider the fact that you have n free choices for what the element 1 goes to, then $n - 1$ remaining choices for the element 2, and so forth until n itself must go to the only element remaining unassigned.   □

In Exercise 8.2 you are asked to investigate the following.

8.2   Proposition: Generators for S_n

The group S_n is generated by the elements $(12), (13), \ldots, (1n)$, which are called *transpositions*.

The elements of S_n are subdivided into *even* and *odd* according to whether they can be written as a product of an even or odd number of transpositions, respectively. For example, $(123) = (12)(13)$ is even and $(1453) = (14)(15)(13)$ is odd. In an abstract algebra course one shows that if a permutation can be written as a product of an even number of transpositions in one way, then all such decompositions for that element contain an even number of transpositions, and similarly for odd permutations. For example,

$$(12)(13)(12)(14) = (14)(23).$$

Clearly, the identity i is even since it can be written either as a product of zero transpositions or in a manner such as $(12)(12)$. In an abstract algebra course one might prove the following results, which I state here without proof. Do note, however, that Proposition 4.18 applies here to establish the normality of A_n.

8.3   Proposition: A_n is a normal subgroup of S_n

The set of even permutations in S_n forms a normal subgroup of S_n, which we call the *alternating group of degree n* and denote by A_n. Here $[S_n : A_n] = 2$ for every $n \geq 2$, and A_n has order $n! \div 2$.

8.4   Theorem: Properties of S_n and A_n

For $n \geq 3$, S_n is nonabelian, and for $n > 3$, A_n is nonabelian. For $n \geq 5$, A_n is simple. The group A_n is generated by the elements (123), (124), . . . , (12n).

Now since D_4 consists of some, but not all of the permutations on the set $\{1, 2, 3, 4\}$, it is certainly a subgroup of S_4. Similarly, the cyclic group of Example 1.3 is a subgroup of S_6 since it may be thought of as six permutations of the vertices of a hexagon. Thus, we formulate the following definition.

8.5   Definition: Permutation group

Any subgroup of S_n is referred to as a *permutation group* on the set $\{1, 2, \ldots, n\}$. The numbers in this set may be referred to as *points* and may be identified with objects such as the vertices of a regular polygon.

It is worthwhile to consider how stabilizers fit into this picture. Let K, for example, be the stabilizer of the point 3 when the group S_5 acts on the set $\{1, 2, 3, 4, 5\}$. Then any permutation on the set $\{1, 2, 4, 5\}$ will be in K, and thus K is isomorphic to S_4.

In a similar sense, nothing prevents us from considering D_4 as a subgroup of S_n for any $n > 4$; we would simply assume that it was a part of the stabilizer of the set $\{5, \ldots, n\}$.

The remainder of this chapter will be devoted to two topics: groups of order $2p$, where p is an odd prime, and the subgroup structure of the group A_5. The first of these is an interesting application of Sylow Theorem III (Theorem 7.13).

8.6   Proposition: Groups of order $2p$

Let p be an odd prime and $|G| = 2p$. Then either $G \cong \mathbf{Z}_{2p}$ or $G \cong D_p$.

Proof: By Theorem 7.13, s_p is a divisor of 2, and is congruent to 1 modulo p. Hence $s_p = 1$ and by Proposition 7.11 G has a normal Sylow p-subgroup P of order p. By Exercise 3.10, P is cyclic and so may be written as $\langle x \rangle$.

Again, by Theorem 7.13, s_2 must be 1 or p. If $s_p = 1$, then G is cyclic exactly as in our discussion of groups of order 35 in Chapter 7. Hence, suppose that $s_p = p$, and let $\langle y \rangle$ be a Sylow 2-subgroup. Now if $\langle z \rangle$ is a Sylow 2-subgroup distinct from $\langle y \rangle$, then $z \neq y$. Thus, G must have p distinct elements of order 2. But G has $p - 1$ distinct elements of order p (namely, $x^2, \ldots, x^{n-1}$) and an identity; thus we have accounted for all the $2p$ elements of G. We thus know that the product xy must either be one of the p elements of order 2 or one of the powers of x. If $xy = x^k$ for some k, then $y = x^{k-1}$. But by Exercise 3.11, the order of x^{k-1} is 1 or p, whereas y has order 2. Hence xy has order 2, that is, $(xy)^2 = 1$, where 1 is the identity of G. But from $xyxy = 1$ together with the fact that $y^{-1} = y$, we have $yx = x^{-1}y^{-1} = x^{-1}y$. Thus, G can be written in terms of the generators and relations in Example 4.4c and so is isomorphic to D_p.   □

Observe that S_3 is nonabelian and of order 6; hence, Proposition 8.6 shows immediately that $S_3 \cong D_3$.

We now turn to the question of determining the subgroups of A_5. This project is quite lengthy, but it will serve to review many of the concepts we have considered thus far and will illustrate how we may apply numerous results to a specific large problem.

By Proposition 8.3, $|A_5| = 60$, so by Theorem 3.10 any subgroup of A_5 must have one of the following orders:

$$1, \ 2, \ 3, \ 4, \ 5, \ 6, \ 10, \ 12, \ 15, \ 20, \ 30, \ 60.$$

Of course, the identity subgroup $\{i\}$ has order 1, and A_5 itself has order 60.

Now, $|A_5| = 60 = 2^2 \cdot 3 \cdot 5$, so by Theorem 7.1 we know that A_5 has subgroups of orders 4, 3, and 5. By Theorem 7.13, $s_5 \equiv 1 \pmod 5$ and s_5 divides 12; hence $s_5 = 1$ or 6. Let's apply some simple combinatorics to determine the number of Sylow 5-subgroups. A 5-cycle such as (12345) is an even permutation and so is in A_5. If we agree to write such a cycle in the form in which 1 stands first (recall that (12345) is the same permutation as (23451), etc.), we then have four choices for the point that stands second, three for the one standing third, two for the one standing fourth, and no choice for the last one. Thus we have accounted for 24 distinct elements of order 5, which determine six distinct subgroups of order 5. By Proposition 4.8 we may treat conjugation as an action on the set of subgroups of A_5; moreover, by Corollary 7.9 the subgroups of order 5 form an orbit under this action. Hence by Theorem 3.13, the stabilizer of one of these subgroups has order $60 \div 6 = 10$. These stabilizers are in fact the normalizers of the six Sylow 5-subgroups, and by Proposition 7.12 these normalizers are distinct. The result is that A_5 has at *least* six subgroups of order 10. If there were any other subgroups of order 10, Theorem 7.1 would apply to them because a subgroup is a group in its own right, and these additional subgroups would likewise have Sylow subgroups of order 5. But there are only six subgroups of order 5 in all of A_5 so there can be only six subgroups of order 10.

That's a slightly daunting paragraph, and you may need to reread it with some care, but the point is that we have determined exactly how many subgroups of orders 5 and 10 there are in A_5. We'll go on to other orders, but let's observe first that this preceding argument actually contained a proof of the following proposition.

8.7 Proposition: The order and number of Sylow p-normalizers

If G is a finite group and if p is a prime dividing $|G|$, then the normalizers of the Sylow p-subgroups of G have order $|G| \div s_p$, and there are exactly s_p such normalizers.

We'll use Proposition 8.7 in the next two steps in our determination of the subgroup structure of A_5.

Let's apply another combinatoric argument to determine the number of 3-cycles in A_5. A 3-cycle will use 3 of the 5 points 1, 2, 3, 4, 5; hence, we must choose 3 of these 5, which can be done in $\binom{5}{3} = 10$ ways. In writing a 3-cycle, we may arbitrarily choose to let the smallest number stand first (since $(abc) = (bca) = (cab)$); then we have two choices as to which point stands second. This gives a total of 20 3-cycles and accounts for 10 cyclic subgroups of order 3; these are the Sylow 3-subgroups of A_5. There cannot be any more because Theorem 7.13 implies that $s_3 = 1$, 4, or 10. By Proposition 8.7, the normalizers of these subgroups have order $60 \div 10 = 6$ and, as before, this must account for all of the subgroups of order 6.

We have accounted thus far for 24 elements of order 5, 20 elements of order 3, and, of course, the identity. This leaves only 15 nonidentity elements in A_5. Another simple combinatorial argument shows that there are already 15 elements that are products of two transpositions such as $(12)(34)$, which means that all of the elements of A_5 for which we have not yet accounted have this form and have order 2. Remember that the Sylow 2-subgroups of A_5 have order 4, not 2. By Theorem 7.13, s_2 is an odd divisor of 15. If s_2 were as small as 3, we would have *at most* 9 elements of order 2, but we have 15; hence, $s_2 = 5$, and there are 5 subgroups of order 4 in A_5. (Since 15 elements of order 2 cannot make 15 subgroups of order

4, we know that $s_2 \neq 15$.) By Proposition 8.7, the normalizers of these subgroups have order $60 \div 5 = 12$; thus there are at least five subgroups of order 12 in A_5.

Thus far the theory of Sylow p-subgroups has enabled us to show that A_5 has subgroups of orders 3, 4, 5, 6, 10, and 12. With the exception of order 12, we also know the exact number of each such subgroup. We'll apply some further considerations to complete the project.

First, we noted above that A_5 has exactly 15 elements of order 2. Each of these, together with the identity i, determines a subgroup of order 2 and exhausts the possibilities for that order.

Now let's think about the stabilizers under A_5 of each of the five points $1, \ldots, 5$. The stabilizer of the point 5 consists of all even permutations on the set $\{1, 2, 3, 4\}$, which is just what we have called A_4. The stabilizer of the point 4 consists of all even permutations on the set $\{1, 2, 3, 5\}$, which differs from $\{1, 2, 3, 4\}$ only in the name given to the last point; thus it is also *an* alternating group of degree 4 and is isomorphic to A_4. In fact, the five stabilizers of the individual points of $\{1, 2, 3, 4, 5\}$ are all isomorphic copies of the abstract group A_4. Thus we know that A_5 has five subgroups of order 12 having this structure. Consideration of the form of the elements in any one of them (even permutations on a set of four elements) shows that they are precisely the Sylow 2-normalizers we found previously.

Are there any other subgroups of order 12 in A_5? This question could be answered by examining conjugates and products in some detail, but we'll appeal instead to a result that is stated here without proof.

8.8   Theorem: The abstract groups of order 12

If a finite group has order 12, then it is isomorphic to one of the following:

(i)   The cyclic group $\mathbf{Z}_{12}$;

(ii)   The dihedral group D_6 of symmetries of the regular hexagon;

(iii) The alternating group A_4 of degree 4;

(iv)   The abelian group given by $\langle a, b : a^6 = b^2 = 1, \ ba = ab \rangle$;

(v)   The generalized quaternion group $Q_3 = \langle x, y : x^6 = 1, \ y^2 = x^3, \ yx = x^{-1}y \rangle$.

The proof of this theorem does not involve difficult or arcane concepts, but it does require a laborious attack by cases. It is not difficult to show that each of the above cases *except* A_4 contains elements of order 6, whereas we know that A_5 has no elements of order 6 (because A_5 has 24 elements of order 5, 20 of order 3, 15 of order 2, the identity, and no others). If A_5 has any subgroup of order 12 other than the five stabilizers we have accounted for, it would have to be isomorphic to A_4. But such a subgroup would itself have a normal subgroup of order 4, and it would be the normalizer of a Sylow 2-subgroup of A_5. What this slightly involved argument has shown is that A_5 has exactly five subgroups of order 12.

Let's consider where we are now. Of the list of possible orders of subgroups for A_5 we have covered every case except orders 15, 20, and 30. We'll show that A_5 has no subgroup of any of these three orders. This will establish the claim made at the beginning of Chapter 7 that the converse of the theorem of Lagrange (Theorem 3.10) is false.

We'll tackle the question of orders 15 and 20 first. Suppose that A_5 has a subgroup H of order 15. In Exercise 7.1 you showed that a group of order 15 must be cyclic; hence, H must contain an element of order 15. But (as above), A_5 has no element of order 15; hence, no such H of order 15 can exist.

Next suppose that A_5 has a subgroup H of order 20. Applying Theorem 7.13 to H, we see that H can have only one Sylow 5-subgroup K. (See Exercise 8.7.) This K is likewise a Sylow 5-subgroup of the whole group A_5, so (since $K \trianglelefteq H$) the subgroup H, which has order 20, would have to be contained in the Sylow 5-normalizer, which has order 10. Thus, A_5 has no subgroup of order 20.

The final case, that of order 30, requires an additional observation.

8.9 Proposition: Orders of elements under homomorphisms

Let G and G^* be finite, $\phi : G \to G^*$ be a homomorphism, and $g \in G$. Then the order of $\phi(g)$ divides *both* the order of the element g and the order of the group G^*. If, further, ϕ is an isomorphism, then the order of $\phi(g)$ equals the order of g. (The proof is left as Exercise 8.11.)

Now suppose that $H \leq A_5$ and that $|H| = 30$. Then by Proposition 4.18, $H \trianglelefteq A_5$. Then, Definition 6.7 gives the canonical homomorphism $\eta : A_5 \to A_5/H$ with $\eta(x) = Hx$ for every $x \in A_5$. Now the elements (123), (124), (125) all have order 3 so by Proposition 8.9, η must carry each of them to an element of A_5/H whose order divides both 3 and also the order of A_5/H, which is 2. Thus, each of these three elements must be carried to the identity H of A_5/H. But from Theorem 8.4 we know that these three elements generate A_5 and since η preserves products, all of the elements of

A_5 are carried to the identity H of A_5/H. Hence the kernel of η is A_5 itself, not H, as is required. This contradiction shows that no such subgroup of order 30 can exist.

Let's conclude this extensive project by describing the structure of the various subgroups of A_5. Those subgroups having prime order must be cyclic. The Sylow 2-subgroups are isomorphic to V_4 (Exercise 8.3). We already discovered that the subgroups of order 12 are all isomorphic copies of A_4. By Proposition 8.6, the subgroups of orders 6 and 10 must be cyclic or dihedral, but cyclic subgroups would require elements of orders 6 and 10, respectively, of which A_5 has none, so these subgroups are dihedral.

This extended discussion constitutes a proof of the following.

8.10   Theorem: Subgroup structure of A_5

The subgroups of A_5 are characterized precisely as follows:

Order	Number of Subgroups	Structure
60	1	Alternating
30	0	
20	0	
15	0	
12	5	Alternating
10	6	Dihedral
6	10	Dihedral
5	6	Cyclic
4	5	Klein-4
3	10	Cyclic
2	15	Cyclic
1	1	Identity

That's really quite a lot of information about A_5.

Exercises

8.1.   Show that consideration of the symmetries of the square with sides a, b, c, d (rather than vertices 1, 2, 3, 4) leads to a group isomorphic to D_4 as given in Example 4.5.

8.2. To see how it happens that the transpositions (12), $(13), \ldots, (1n)$ generate S_n, first show that an arbitrary transposition $(\alpha\beta)$, where α and β are distinct members of the set $\{1, 2, \ldots, n\}$, can be written as the product of *three* transpositions of the form (1γ), where in each of the three, γ is either α or β. Products like $(\alpha\beta)(\gamma\delta)$ with distinct $(\alpha, \beta, \gamma, \delta)$ provide additional elements of S_n. Now mutiply out the sample products $(\alpha\beta)(\alpha\gamma)$ and $(\alpha\beta)(\alpha\gamma)(\alpha\delta)$ to see what elements you get. This much does not constitute a *proof* of Proposition 8.2, but it should give you an idea of how general elements arise as products of the generators.

8.3. Show that the Sylow 2-subgroups of A_5 are isomorphic to V_4.

8.4. Find an element of order 6 in each of the abstract groups of order 12 except for A_4.

8.5. The group G having the presentation $\langle p, q, r : p^3 = q^2 = r^2 = 1, pq = qp, pr = rp, qr = rq \rangle$ has order 12. Find to which of the groups in Theorem 8.8 G is isomorphic.

8.6. List the elements of the group in Theorem 8.8(iv) and find the order of each element. If σ is a homomorphism from this group into a group of order 4, which elements must necessarily be in the kernel of σ? Can σ map this group *onto* Z_4?

8.7. Show that A_5 has no subgroup of order 15 by showing that such a subgroup would have a normal Sylow 5-subgroup. Explain why this leads to a contradiction.

8.8. Determine the subgroup structure of A_4 by reasoning similar to that used for A_5.

8.9. Determine the order and number of the Sylow 3- and 5-normalizers of S_5. (Be sure to check *all* possible values of s_3.)

8.10. Determine the order and number of the Sylow 2-normalizers of S_5.

8.11. Prove Proposition 8.9.

8.12. Let G and G^* be finite, $\phi : G \to G^*$ be an isomorphism, and $g \in G$ have order k. Prove that k divides the order of G^*.

8.13. Prove that there are only two abstract groups of order 4. (Note that this extends the result of Proposition 8.6 to the even prime 2. It also supplies an alternative approach to Exercise 8.3.)

8.14. Find the normalizers of the Sylow 3-subgroups of A_4.

8.15. A Sylow 3-subgroup of A_5 has the form $P = \{e, (abc), (acb)\}$, where a, b, and c are distinct elements of the set $\{1, 2, 3, 4, 5\}$. In terms of these variables, find the Sylow normalizer of P. If Q is a Sylow 3-subgroup of A_5 distinct from P, find the intersection of the normalizers of the two subgroups.

Chapter 9

Matrix Groups

In Example 1.1 we considered a group consisting of six matrices under the operation of matrix multiplication. Since Example 1.3 was concerned with rotations of a regular hexagon in the plane, we can also express this example in terms of matrices. You should recall from linear algebra that a rotation counterclockwise about the origin through an angle θ may be expressed as a linear transformation with corresponding matrix:

$$\begin{bmatrix} \cos\theta & -\sin\theta \\ \sin\theta & \cos\theta \end{bmatrix}.$$

This matrix, applied to the column vector $\begin{bmatrix} x \\ y \end{bmatrix}$ gives the vector resulting from a rotation through angle θ counterclockwise. In Example 1.3, $\theta = 60° = \pi/3$, so we may position the hexagon with its center at the origin of the xy-plane and identify the rotation $\mathbf{R}$ with the matrix:

$$\mathbf{R} = \begin{bmatrix} \frac{1}{2} & -\frac{\sqrt{3}}{2} \\ \frac{\sqrt{3}}{2} & \frac{1}{2} \end{bmatrix}.$$

You should check that this $\mathbf{R}^6 = I$, as required. Note, in particular, that $\mathbf{R}^3 = -I$, which represents a rotation through $180°$.

The group of the square may similarly be represented as the group generated by the matrices:

$$r = \begin{bmatrix} 0 & -1 \\ 1 & 0 \end{bmatrix} \quad \text{and} \quad c = \begin{bmatrix} 1 & 0 \\ 0 & -1 \end{bmatrix},$$

where c denotes a reflection in the horizontal axis through the center of the square. Exercise 9.1 asks you to verify that the relations given in Example 4.5 are satisfied.

It is worth remarking that the $m \times n$ matrices (for fixed m and n) under the operation of matrix addition also form a group, but we will not have occasion to use such an additive group in this book. Of course, in working with the operation of multiplication we shall have to use square matrices of a given size.

We'll also have occasion to distinguish between matrices whose entries come from the field **R** of real numbers and those with entries from the field **C** of complex numbers. Recall that a square matrix has a multiplicative inverse if and only if it is *nonsingular*, that is, it has a nonzero determinant.

9.1   Definition: The groups $GL(n, \mathrm{R})$ and $GL(n, \mathrm{C})$

For any positive integer n, the set of nonsingular $n \times n$ matrices with entries from the real numbers **R** under the operation of matrix multiplication forms the *general linear group* $GL(n, \mathbf{R})$; those with entries from the complex numbers **C** form the *general linear group* $GL(n, \mathbf{C})$.

The basic properties of nonsingular matrices of a given size ensure that the sets above are in fact groups.

In the discussion above, the matrices **R**, r, and c could be considered as members either of $GL(2, \mathbf{R})$ or of $GL(2, \mathbf{C})$. However,

$$Z = \begin{bmatrix} 0 & i \\ 1 & 0 \end{bmatrix}$$

must be thought of as an element of $GL(2, \mathbf{C})$. In that group, the powers of Z form a cyclic subgroup; see Exercise 9.2.

For the moment, let K stand for either **R** or **C**. Two subgroups of $GL(n, K)$ are important for applications; they are defined as follows.

9.2   Definition: The special linear (unimodular) groups

The matrices having determinant 1 form a subgroup $SL(n, K)$ of $GL(n, K)$ called the *special linear group over K*; this subgroup is also called the *unimodular group*.

The next definition is parallel to Definition 8.5.

9.3 Definition: Matrix group

Any subgroup of $GL(n, \mathbf{C})$ is called a *matrix group*.

A matrix group may be finite or infinite; see Exercises 9.2, 9.6, and 9.7. A useful example of an infinite cyclic matrix group is $\langle X \rangle$, where

$$X = \begin{bmatrix} 1 & 1 \\ 0 & 1 \end{bmatrix} \qquad \text{and} \qquad X^n = \begin{bmatrix} 1 & n \\ 0 & 1 \end{bmatrix}$$

for every integer n (positive, negative, or 0).

From linear algebra, you know that a matrix corresponds to a linear transformation of a vector space (to itself or to another vector space). Thus, a matrix group corresponds to a group of linear transformations (under the operation of composition of functions) on a vector space, and we'll make this identification in the chapters that follow. The concept of matrix groups gives rise to the entire theory of group representations and characters.

We'll conclude this chapter with one further idea and comment.

9.4 Definition: Primitive nth root of 1

A complex number ζ is a *primitive nth root of unity* if $\zeta^n = 1$ and if $\zeta^m \neq 1$ for $1 \leq m < n$.

For example, both i and $-i$ are primitive fourth roots of 1; 1 and -1 are fourth roots of 1, but not primitive. Both

$$\frac{-1 + i\sqrt{3}}{2} \qquad \text{and} \qquad \frac{-1 - i\sqrt{3}}{2}$$

are primitive third roots of 1. If ζ is a primitive nth root of 1, then the 1×1 matrix $[\zeta]$ generates a cyclic group of order n. Although the 1×1 matrices comprising $GL(1, \mathbf{C})$ are of little (if any) consequence in linear algebra, they are of great importance in group representation theory.

Exercises

9.1. Verify that the matrices given in the text for the symmetries r and c of the square satisfy the relations given in Example 4.5 for the group D_4.

9.2. Consider the matrices

$$Z = \begin{bmatrix} 0 & i \\ 1 & 0 \end{bmatrix} \quad \text{and} \quad C = \begin{bmatrix} 1 & i \\ 0 & 1 \end{bmatrix};$$

show that $\langle Z \rangle$ is a finite cyclic subgroup of $GL(2, \mathbf{C})$ but that $\langle C \rangle$ is infinite cyclic. What is C^{-1}?

9.3. Show that for $n = 1$, $GL(n, \mathbf{R}) \trianglelefteq GL(n, \mathbf{C})$. Then show that normality does not hold for $n = 2$ by considering $Y^{-1}XY$, where

$$X = \begin{bmatrix} 1 & 2 \\ 3 & 1 \end{bmatrix} \quad \text{and} \quad Y = \begin{bmatrix} 1 & i \\ 0 & 1 \end{bmatrix}.$$

9.4. Prove that $SL(n, \mathbf{C}) \trianglelefteq GL(n, \mathbf{C})$. (Of course, $SL(n, \mathbf{R}) \trianglelefteq GL(n, \mathbf{R})$ also.)

9.5. Let ζ be a (complex) primitive kth root of 1, and let G consist of the matrices in $GL(n, \mathbf{C})$ whose determinants are powers of ζ. Prove that $G \trianglelefteq GL(n, \mathbf{C})$. (Don't forget to show that G is a subgroup or to show that it is normal.)

9.6. Let

$$A = \begin{bmatrix} 0 & 1 & 0 \\ 0 & 0 & 1 \\ 1 & 0 & 0 \end{bmatrix} \quad \text{and} \quad B = \begin{bmatrix} 0 & -1 & 0 \\ 0 & 0 & 1 \\ 1 & 0 & 0 \end{bmatrix}.$$

Show that A and B generate cyclic groups of order 3 and 6, respectively.

9.7. (Same as Exercise 1.7) Let

$$A = \begin{bmatrix} 0 & 1 \\ 1 & 0 \end{bmatrix} \quad \text{and} \quad B = \begin{bmatrix} -1 & 0 \\ 0 & -1 \end{bmatrix};$$

show that $\langle A, B \rangle$ is isomorphic to V_4.

9.8. Suppose that $\theta = 2\pi/r$, where r is an *irrational* number, and that

$$R = \begin{bmatrix} \cos\theta & -\sin\theta \\ \sin\theta & \cos\theta \end{bmatrix}.$$

Show that $\langle R \rangle$ is an infinite cyclic matrix group of rotations.

Chapter 10

Group Representations

In the preceding chapter we had some examples in which groups that were already familiar were identified with matrix groups. In particular, Exercises 9.1, 9.2, 9.6, and 9.7 set up isomorphisms between abstract groups and matrix groups. Such correspondences turn out to be useful in group theory, but we do not need to limit our consideration to isomorphisms. Hence, we'll introduce the following terminology.

10.1 Definition: Group representation

Let G be any arbitrary group and G^* a matrix group, that is, a subgroup of $GL(n, \mathbf{C})$. A homomorphism $T : G \to G^*$ is called a *representation of* G. The integer n is called the *degree* or *dimension* of T and is denoted by $\deg(T)$.

The effect of a representation is to replace the elements of G by nonsingular $n \times n$ matrices and to replace the group operation with matrix multiplication. For example, let G be the cyclic group of order 2, written as $\{1, x\}$. Then,

$$T(x) = \begin{bmatrix} 0 & 1 \\ 1 & 0 \end{bmatrix}, \qquad T(1) = \begin{bmatrix} 1 & 0 \\ 0 & 1 \end{bmatrix}$$

is a representation of G; the dimension of T is 2. For a one-dimensional representation of G, we could consider

$$U(x) = [-1], \qquad U(1) = [1].$$

Note that a one-dimensional representation is a function whose values are

1×1 matrices, which are distinguished in form (if not in substance) from real or complex numbers as such.

Of course, each of the above examples is an isomorphism, but we have not required a representation to be $1:1$. For example, write $Z_4 = \{1, a, a^2, a^3\}$ and set

$$T(a) = \begin{bmatrix} 0 & 1 \\ 1 & 0 \end{bmatrix};$$

then the powers of $T(a)$ will constitute the rest of the matrices representing Z_4. Since $T(a)^2 = I_2$, the identity matrix, T is a representation but not an isomorphism. This distinction is made explicit in the following.

10.2   Definition: Faithful representation

A representation T of a group G is called *faithful* if T is a $1:1$ function on G.

The correspondences set up in Exercises 9.1, 9.2, 9.6, and 9.7 are all faithful representations.

We have defined a representation to be a particular kind of function (namely, an operation-preserving one) from a group in which we are interested into a group of nonsingular matrices. The value of representations is primarily that computations with matrices are relatively easy to handle (whereas group operations in general can be messy) and that many theorems about matrices illuminate the theory of groups by means of representations.

As previously noted, if a group G is specified in terms of generators and relations, it suffices to define a representation T on the generators of G since all other values of T will be determined by the requirement that a homomorphism preserves products and powers (which are, of course, themselves products). In particular, such a T must preserve inverses in the sense that $T(g^{-1}) = (T(g))^{-1}$, and T must carry the identity of G to the $n \times n$ identity matrix.

Recall from Exercise 8.12 that there are exactly two abstract groups of order 4, namely, Z_4 and V_4. Let $G = \langle x \rangle$ be cyclic of order 4, and set

$$T(x) = \begin{bmatrix} 1 & 0 & 0 \\ 0 & 0 & -1 \\ 0 & 1 & 0 \end{bmatrix}.$$

Since $T(x)^4 = I_3$, the defining relation $x^4 = 1$ of G is preserved by T, and T is a representation of G. For the same group G, let

$$U(x) = \begin{bmatrix} 0 & 1 \\ 1 & 0 \end{bmatrix};$$

then U is also a representation of G, although U is not faithful, whereas T is.

Exercise 9.7 gave a faithful representation for $V_4 = \{1, a, b, ab\}$. Let

$$U(a) = U(b) = \begin{bmatrix} 0 & 1 \\ 1 & 0 \end{bmatrix};$$

then U is a homomorphism with kernel $\{1, ab\}$ and so is a representation, but is not faithful.

Although the notion of a 1×1 matrix may seem slightly contrived to you, the representations of degree 1 turn out to be important, so let's find all of the one-dimensional representations of the nonabelian group $D_3 = \langle r, c \rangle$ of order 6. (By Proposition 8.6 these will also be all of the one-dimensional representations of the isomorphic group S_3.) If T is a one-dimensional representation of D_3, then $T(c)$ must be a 1×1 matrix whose square is the identity matrix $[1]$; hence $T(c) = [\pm 1]$. Similarly, if $T(r) = [\zeta]$, then we need $T(r)^3 = [1]$, so ζ must be a cube root of 1. Suppose we take ζ to be a primitive cube root, that is,

$$\zeta = -\frac{1}{2} \pm i\frac{\sqrt{3}}{2}.$$

Here, if we choose for ζ the plus sign from the $\pm$, ζ^2 will have the minus sign, and vice versa. Now if we take $T(r) = [\zeta]$ and $T(c) = [1]$, then $T(cr) = [\zeta]$ but $T(r^2c) = [\zeta^2]$, so the relation $cr = r^2c$ is not preserved by T, and T is not a homomorphism. Similarly, if we take $T(r) = [\zeta]$ and $T(c) = [-1]$, then $T(cr) \neq T(r^2c)$. Thus the only choice for $T(r)$ is $[1]$. Now let's take

$$T_1(r) = [1] \quad \text{and} \quad T_1(c) = [1],$$
$$T_2(r) = [1] \quad \text{and} \quad T_2(c) = [-1].$$

Both of these preserve the relation $cr = r^2c$ and so are representations of D_3. In fact, our discussion has shown that these are the *only* one-dimensional representations of D_3.

In this example, T_1 merely assigns the matrix $[1]$ to every element of D_3, which may seem artificial to you, but this function certainly satisfied Definition 10.1. More generally, for an arbitrary group G and degree n, the function

$$T(g) = I_n \quad \text{for every } g \in G$$

is a representation of G. This T is called the *trivial representation of degree* n, but the triviality is in the construction (or the omnipresence) of this representation — you will see that it is not trivial in the sense of being unimportant!

Now an $n \times n$ matrix A determines a linear transformation on a vector space of dimension n. Specifically, if $A \in GL(n, \mathbf{C})$, then the linear transformation given by A is one on the space $\mathbf{C}^n$ of n-dimensional vectors with complex entries. If T is a representation of degree n, we say that $\mathbf{C}^n$ *affords* T, or that T is a representation *over* $\mathbf{C}$. In the special case in which all $T(g)$ have only real entries, we say that $\mathbf{R}^n$ affords T, or that T is a representation over $\mathbf{R}$.

By considering the Euclidean spaces $\mathbf{R}^2$ and $\mathbf{R}^3$, we can use our geometric examples from the plane and from three-dimensional space to obtain matrix representations of the abstract groups to which they are isomorphic. First, however, some comment on notation will be needed. If X is an n-dimensional *column* vector and A an $n \times n$ matrix, then from linear algebra you should recall that AX is the vector to which the transformation determined by A carries X. We'll use the superscript t to denote transpose, and recall that the transpose of a column vector is a row vector. From linear algebra,

$$(AX)^t = X^t A^t;$$

hence $X^t A^t$ is the row vector to which the transformation determined by A^t carries X^t. You most likely used column vectors when you studied linear algebra; in representation theory it is more convenient to employ row vectors for two reasons. The first and less important is that it maintains consistency with our notation in Definition 3.1 for a group action on a point set. The more crucial consideration is that we have interpreted the product xy for x and y in a group to mean "x followed by y"; thus if T is a representation of G, we want $T(xy) = T(x)T(y)$ with the right-hand side of this equation meaning "$T(x)$ followed by $T(y)$." If we use *column* vectors X, we must define a representation T to be an *antihomomorphism*; that is, we would take $T(xy) = T(y)T(x)$ so that $T(xy)X = T(y)(T(x)X)$, but if we use *row* vectors Y, then

$$Y\,T(xy) = Y\,T(x)T(y)$$

is consistent with our definition of a representation as a homomorphism.

The representation derived in Chapter 9 for D_4 referred to column vectors, but it can easily be modified to illustrate the use of row vectors. Specifically, for

$$T(r) = \begin{bmatrix} 0 & 1 \\ -1 & 0 \end{bmatrix} \quad \text{and} \quad T(c) = \begin{bmatrix} 1 & 0 \\ 0 & -1 \end{bmatrix},$$

then row vector (x, y) represents a point in $\mathbf{R}^2$ and

$$(x, y)T(r) = (x, y) \begin{bmatrix} 0 & 1 \\ -1 & 0 \end{bmatrix} = (-y, x)$$

is a rotation through 90° counterclockwise, while

$$(x, y)T(c) = (x, y) \begin{bmatrix} 1 & 0 \\ 0 & -1 \end{bmatrix} = (x, -y)$$

is a reflection in the horizontal axis. Since the linear transformations correspond to the motions described, it is clear that T is a faithful representation of D_4.

Since the group of the square bipyramid is isomorphic to D_4, we may consider its geometry in $\mathbf{R}^3$ in order to find another representation of D_4. Place the solid as shown in Figure 5.1 (Chapter 5), and take $V(r)$ to be a rotation through 90° about the z-axis. We can write this in matrix terms as

$$(x, y, z)V(r) = (x, y, z) \begin{bmatrix} 0 & 1 & 0 \\ -1 & 0 & 0 \\ 0 & 0 & 1 \end{bmatrix} = (-y, x, z).$$

Then we may take $V(c)$ as a rotation through 180° about the x-axis, writing

$$(x, y, z)V(c) = (x, y, z) \begin{bmatrix} 1 & 0 & 0 \\ 0 & -1 & 0 \\ 0 & 0 & -1 \end{bmatrix} = (x, -y, -z).$$

Then V determines a representation of degree 3 for D_4. Note that the matrices $V(r)$ and $V(c)$ may each be partitioned as

$$\begin{bmatrix} a_{11} & a_{12} & 0 \\ a_{21} & a_{22} & 0 \\ \hline 0 & 0 & a_{33} \end{bmatrix}.$$

The product of any two matrices of this form again has this form (see Exercise 10.1). More specifically, if we let $U(r) = [1]$ and $U(c) = [-1]$, then

$$V(r) = \begin{bmatrix} & & 0 \\ T(r) & & 0 \\ & & 0 \\ \hline 0 & 0 & U(r) \end{bmatrix}, \qquad V(c) = \begin{bmatrix} & & 0 \\ T(c) & & 0 \\ & & 0 \\ \hline 0 & 0 & U(c) \end{bmatrix},$$

and by virtue of Exercise 10.1, we have

$$
V(g) = \left[
\begin{array}{cc|c}
 & & 0 \\
\multicolumn{2}{c|}{T(g)} & \\
 & & 0 \\
\hline
0 & 0 & U(g)
\end{array}
\right]
$$

for every $g \in D_4$. You can see that V decomposes into two smaller representations T and U. Conversely, if we have representations T and U of a group G, with m the degree of T and n the degree of U, then

$$
V(g) = \begin{bmatrix} T(g) & \mathbf{0} \\ \mathbf{0} & U(g) \end{bmatrix} \qquad \text{for each } g \in G,
$$

defines a representation of G, as is easy to see. Here the boldface $\mathbf{0}$ denotes an $m \times n$ submatrix of zeros in the upper right-hand corner and an $n \times m$ submatrix of zeros in the lower left-hand corner. We'll use this notation $\mathbf{0}$ frequently with the understanding that the submatrix of zeros always has the proper number of rows and columns to make the large matrix square.

Note that the placement of the intersection of the axes of symmetry of the solid at the origin in Figure 5.1 is essential since a linear transformation always carries the origin of $\mathbf{R}^n$ to itself.

For another example, recall from Application 3.14 the group G of rigid symmetries of the cube. This time we'll take r to be a *clockwise* rotation (when seen from above) through $90°$ about the z-axis. Then r may be represented by the cube of the matrix $V(r)$ from the preceding example (why?), and we set

$$
T(r) = \begin{bmatrix} 0 & -1 & 0 \\ 1 & 0 & 0 \\ 0 & 0 & 1 \end{bmatrix}.
$$

For simplicity, we'll take the edges of the cube to have length 2; then in Figure 10.1 we have the following assignment:

Vertex	Coordinates
1	$(1, -1, 1)$
2	$(-1, -1, 1)$
3	$(-1, 1, 1)$
4	$(1, 1, 1)$
5	$(1, -1, -1)$
6	$(-1, -1, -1)$
7	$(-1, 1, -1)$
8	$(1, 1, -1)$

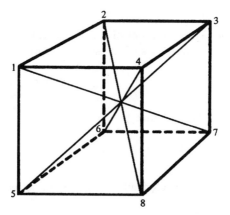

FIGURE 10.1
Cube with axes

Then in terms of the coordinates of the eight vertices, r is the permutation $(1234)(5678)$. Let $d = (245)(386)$, a rotation of $120°$ about the diagonal joining vertices 1 and 7. Then d is described by the linear transformation

$$T(d) = \begin{bmatrix} 0 & -1 & 0 \\ 0 & 0 & -1 \\ 1 & 0 & 0 \end{bmatrix}.$$

With a little patience you can write down the 24 elements of G in terms of r and d; hence, to define T on the rest of G we merely extend by powers and products as usual. It remains to show that T is indeed a homomorphism on G, but this fact is tedious to verify by calculation and easy to check by observation. After all, T is merely a specification of the linear transformations performed by the elements of G; thus the set of matrices

$$\{T(g) : g \in G\}$$

forms a group that is isomorphic to G in the sense of being the same elements in a different notation. Equivalently, this set forms the same abstract group as does G.

Although in this example the matrix $T(r)$ has the form

$$\begin{bmatrix} A(r) & \mathbf{0} \\ \mathbf{0} & B(r) \end{bmatrix}$$

with $A(r)$ 2×2 and $B(r)$ 1×1, $T(d)$ does not have this form. We'll see later that it cannot be rewritten in such a form.

To introduce the last main point of this chapter, let's return to the group D_4, regarded as the symmetries of the square. If we write

$$D_4 = \langle r, c : r^4 = c^2 = 1, cr = r^3 c \rangle$$

and then set $b = cr$, we find that $b^2 = 1$ and that

$$br = (cr)r = (r^3 c)r = r^3 b,$$

and this has given us the reformulation of D_4 in terms of generators and relations as

$$D_4 = \langle r, b : r^4 = b^2 = 1, br = r^3 b \rangle.$$

It should be clear that $\sigma(r) = r$, $\sigma(c) = b$ specifies an automorphism of D_4. (Refer to Exercise 5.14.) Now in terms of the representation T previously constructed,

$$T(b) = T(cr) = T(c)T(r) = \begin{bmatrix} 0 & 1 \\ 1 & 0 \end{bmatrix}.$$

In view of the isomorphic presentations for D_4 as $\langle r, c \rangle$ and $\langle r, b \rangle$, we'll define another representation T^* of D_4 by $T^*(r) = T(r)$, $T^*(c) = T(b)$, bearing in mind, as usual, that the remaining values of T^* on D_4 are formed as products of the given matrices. You can check that T^* is indeed a representation by verifying that

$$T^*(r)^4 = T^*(c)^2 = I_2, \qquad T^*(c)T^*(r) = T^*(r)^3 T^*(c).$$

Now T and T^* are different functions (they take different values on c, among other elements), but they were produced from a single geometric interpretation, in which b merely represents a reflection in the line $x = y$, by expoiting the automorphism σ specified above. Thus in a very concrete sense, T and T^* are the same representation — they amount to exactly the same set of eight linear transformations of $\mathbf{R}^2$. We need an explicit formulation of the statement that two representations are essentially the same, even though they are not identical as functions.

The idea of equivalence of representations is derived from that of linear transformations with a change of basis. Recall that two linear transformations L and L^* on a vector space V are *equivalent* if there is a change of basis for which (using composition of functions) $L^*B = BL$; that is,

$$L^*(B(v)) = B(L(v)) \qquad \text{for every } v \in V.$$

In terms of matrices then, two matrices M and M^* are *equivalent* if there is a nonsingular matrix A such that $M^*A = AM$, or what is the same thing, if $A^{-1}M^*A = M$. If, for column vectors X, we write $L(X) = MX$ and $L^*(X) = M^*X$, the two ideas of equivalence are the same. Here the nonsingular matrix A represents the change-of-basis B in the vector space V. To apply these ideas to representation theory, we formulate a definition.

10.3 Definition: Equivalent representations

Let T and U be n-dimensional representations of a group G. We say that T and U are *equivalent* if there is a nonsingular matrix A such that $AT(g) = U(g)A$ for *every* $g \in G$.

It is absolutely essential to bear in mind that *the same matrix A* must work simultaneously *for every* $g \in G$; thus the matrix A must be independent of the element g in the equation $AT(g) = U(g)A$. Moreover, do not forget that A must be *nonsingular*.

We'll illustrate Definition 10.3 by returning to the two representations T and T^* that we found for D_4. To show that T and T^* are equivalent, we need a matrix

$$A = \begin{bmatrix} a & b \\ c & d \end{bmatrix}$$

with $\det(A) = ad - bc \neq 0$ such that $AT(g) = T^*(g)A$ for *every* $g \in D_4$. In particular, we need $AT(r) = T^*(r)A$, so we require

$$\begin{bmatrix} a & b \\ c & d \end{bmatrix} \begin{bmatrix} 0 & 1 \\ -1 & 0 \end{bmatrix} = \begin{bmatrix} 0 & 1 \\ -1 & 0 \end{bmatrix} \begin{bmatrix} a & b \\ c & d \end{bmatrix};$$

that is,

$$\begin{bmatrix} -b & a \\ -d & c \end{bmatrix} = \begin{bmatrix} c & d \\ -a & -b \end{bmatrix},$$

which gives $a = d$ and $b = -c$. Moreover, we require that $AT(c) = T^*(c)A$, so

$$\begin{bmatrix} a & b \\ c & d \end{bmatrix} \begin{bmatrix} 1 & 0 \\ 0 & -1 \end{bmatrix} = \begin{bmatrix} 0 & 1 \\ 1 & 0 \end{bmatrix} \begin{bmatrix} a & b \\ c & d \end{bmatrix};$$

that is,

$$\begin{bmatrix} a & -b \\ c & -d \end{bmatrix} = \begin{bmatrix} c & d \\ a & b \end{bmatrix},$$

which yields $a = c$ and $b = -d$. Now for any constant $\lambda \neq 0$, $AT(g) = T^*(g)A$ if and only if $(\lambda A)T(g) = T^*(g)(\lambda A)$, so we have a free choice of one nonzero entry in A, say, $a = 1$. Then

$$A = \begin{bmatrix} 1 & -1 \\ 1 & 1 \end{bmatrix},$$

and you can easily verify that $AT(g) = T^*(g)A$ for every $g \in D_4$. Instead of making the arbitrary choice $a = 1$, we could have chosen $a = 1/\sqrt{2}$ to make A orthogonal.

Fortunately, you do not have to check *every* equation of the form $AT(g) = U(g)A$ to show that representations T and U are equivalent, as in Definition 10.3. It suffices to check this equality for the elements of a set of generators, as follows.

10.4   Proposition: Equivalence may be checked on a generating set

Let T and U be representations of a group G and let G be generated by $x_1, \ldots, x_n$. If there is a nonsingular matrix A such that $AT(x_i) = U(x_i)A$ for all $i = 1, \ldots, n$, then T is equivalent to U.

Proof: The proof is left as Exercises 10.11 and 10.12.

Exercises

10.1.   Consider a collection of matrices of the form:

$$\begin{bmatrix} A & 0 \\ 0 & B \end{bmatrix}$$

where A is an $n \times n$ matrix, B is an $m \times m$ matrix, and the boldface **0** terms represent rectangular blocks of zeros as described in the text. Prove that the product of any two such matrices has the same form. (If such a matrix is nonsingular, its inverse has this same form.)

10.2.   Find all of the one-dimensional representations of $V_4 = \{1, a, b, ab\}$.

10.3.   Let V be the three-dimensional representation of the square-bipyramid group given in the text. Find $V(g)$ for each element of the group.

10.4. Given that

$$T(r) = \begin{bmatrix} 0 & -1 \\ 1 & 0 \end{bmatrix}, \qquad T(c) = \begin{bmatrix} 0 & 1 \\ 1 & 0 \end{bmatrix}$$

and

$$U(r) = \begin{bmatrix} 0 & -1 & 0 & 0 \\ 1 & 0 & 0 & 0 \\ 0 & 0 & 0 & -1 \\ 0 & 0 & 1 & 0 \end{bmatrix}, \qquad U(c) = \begin{bmatrix} 0 & 1 & 1 & 0 \\ 1 & 0 & 0 & -1 \\ 0 & 0 & 0 & 1 \\ 0 & 0 & 1 & 0 \end{bmatrix}$$

both determine representations for $D_4 = \langle r, c \rangle$, prove that U is equivalent to the representation

$$V(g) = \begin{bmatrix} T(g) & \mathbf{0} \\ \mathbf{0} & T(g) \end{bmatrix} \qquad \text{for } g \in D_4.$$

10.5. Consider $D_3 = \langle r, c \rangle$ as the symmetries of an equilateral triangle, placed in the plane so that one vertex is at the point $(1, 0)$ and the other two vertices are in the second and third quadrants. Find a two-dimensional representation of D_3 for which $T(r)$ is a rotation through $120°$ counterclockwise and $T(c)$ is a flip in the x-axis. Check that $T(c)T(r) = T(r)^2 T(c)$.

10.6. Let $T_1(r) = [1]$, $T_1(c) = [-1]$ determine a representation for D_3, and let T_2 be the trivial one-dimensional representation of D_3. Prove that the representation found in Exercise 10.5 is *not* equivalent to

$$V(g) = \begin{bmatrix} T_1(g) & \mathbf{0} \\ \mathbf{0} & T_2(g) \end{bmatrix}, \qquad \text{for } g \in D_3.$$

10.7. Show that

$$T(g) = \begin{bmatrix} 0 & 1 \\ -1 & 0 \end{bmatrix}$$

determines a representation of the cyclic group $\langle g : g^4 = 1 \rangle$ corresponding to a rotation of $90°$ in the plane. Then show that $U(g) = [-i]$ and $U^*(g) = [i]$ also determine representations of Z_4. Finally, show that T is equivalent to the representation given by

$$V(g) = \begin{bmatrix} U(g) & 0 \\ 0 & U^*(g) \end{bmatrix}.$$

10.8.   Find all of the one-dimensional representations of D_4.

10.9.   Find all of the one-dimensional representations of Q_2.

10.10. Suppose that two *one-dimensional* representations T and U are equivalent. What can you say about T and U?

10.11. Let T and U be representations of G, $g \in G$, and A be a nonsingular matrix such that $AT(g) = U(g)A$ (for the particular element g). Show that $AT(g^{-1}) = U(g^{-1})A$ as well.

10.12. Using Exercise 10.11, complete the proof of Proposition 10.4.

Chapter 11

Regular Representations

In this chapter we'll meet one of the most important representations of a finite group. Let's begin by considering a new action of a group on the set of its own elements.

11.1 Proposition: Right multiplication is an action

Let G be any group and for any x, $g \in G$, let $g^x = gx$; we say that x acts on the elements of G by right multiplication. Then this operation is an action of the group G on the set G.

Proof: By the closure property in a group, g^x is always defined. To check the conditions of the definition of an action in Definition 3.1, observe that

$$(g^x)^y = (gx)^y = (gx)y = g(xy) = g^{xy}$$
$$\text{and} \quad g^1 = g,$$

where 1 is the identity element of G. $\qquad \square$

It will be convenient in the subsequent discussion for us to have the alternative of writing this action of right multiplication in functional notation as

$$x_R : g \to gx \quad \text{for } g \in G.$$

For the action of right multiplication, the orbit of any element $g \in G$ is all of G; that is, $g^G = G$ in the notation of Definition 3.2 of orbit. This fact follows directly from Proposition 2.12. Correspondingly, the stabilizer of an element g is merely the identity subgroup $\{1\}$ since $g^x = gx = g$ only when $x = 1$.

Now let G be a finite group of order n and write

$$G = \{x_1, x_2, \ldots, x_n\}.$$

We'll consider the action of a *fixed* element x_i on the set G (where, of course, $1 \le i \le n$). The result is a permutation of the elements x_j among themselves; for each j with $1 \le j \le n$, there exists a unique k with $1 \le k \le n$, such that

$$x_j x_i = x_k \qquad \text{or equivalently} \qquad (x_i)_R : x_j \to x_k.$$

We may rewrite the above as $x_i = x_j^{-1} x_k$.

Now let $\mathbf{R}^n$ denote n-dimensional Euclidean space. To each x_j of G with $1 \le j \le n$, we associate the unit *row* vector e_j having 1 as its jth component and 0 for each of the remaining components. Then, under this correspondence, right multiplication gives the linear transformation determined by

$$(x_i)_R : e_j \to e_k$$

for the unique k that we associated with each j in writing $x_j x_i = x_k$. This transformation has associated with it a matrix M which has a 1 in its (j, k)-entry if $(x_i)_R : e_j \to e_k$ and 0 in all other entries. In these two paragraphs we have held x_i fixed but arbitrary in G; of course, each action $(x_1)_R, (x_2)_R, \ldots, (x_n)_R$ induces a linear transformation on $\mathbf{R}^n$.

We'll give a formal definition for the right regular representation shortly, but let's start with two examples, a cyclic group of order 3 and a nonabelian group.

11.2   Example: Right regular representation of Z_3

Let $G = \{1, g, g^2\}$, and write $x_1 = 1$, $x_2 = g$, $x_3 = g^2$. Then we have

$$(x_1)_R = 1_R : e_j \to e_j \qquad\qquad \text{for } j = 1, 2, 3;$$
$$(x_2)_R = g_R : \begin{cases} e_j \to e_{j+1} \\ e_3 \to e_1; \end{cases} \qquad \text{for } j = 1, 2;$$
$$(x_3)_R = g_R^2 : \begin{cases} e_1 \to e_3, \\ e_j \to e_{j-1} \end{cases} \qquad \text{for } j = 2, 3.$$

Hence the matrix $T(g)$ representing the linear transformation g_R must satisfy

$$e_1 T(g) = e_2, \qquad e_2 T(g) = e_3, \qquad e_3 T(g) = e_1,$$

where e_1, e_2, e_3 are *row* vectors. Therefore, $T(g)$ must be the matrix

$$T(g) = \begin{bmatrix} 0 & 1 & 0 \\ 0 & 0 & 1 \\ 1 & 0 & 0 \end{bmatrix},$$

and by similar reasoning,

$$T(g^2) = \begin{bmatrix} 0 & 0 & 1 \\ 1 & 0 & 0 \\ 0 & 1 & 0 \end{bmatrix}$$

and $T(1) = I_3$, the identity matrix. You should check that $T(g)^2 = T(g^2)$ and $T(g)^3 = I_3$, showing that T is indeed a representation of this cyclic group of order 3.

We'll look next at a nonabelian example, but you should have observed by now that the right regular representation has degree equal to the order of the group. So to keep the matrices as manageable as possible, let's take the *smallest* nonabelian group, which is D_3, of order 6.

11.3 Example: Right regular representation of D_3

We'll use the familiar dihedral notation $D_3 = \langle r, c : r^3 = c^2 = 1, cr = r^2 c \rangle$ rather than $x_1, \ldots, x_6$. Then,

$$
\begin{array}{lll}
r_R : 1 \rightarrow r, & r_R : r \rightarrow r^2, & r_R : r^2 \rightarrow 1, \\
r_R : c \rightarrow r^2 c, & r_R : rc \rightarrow c, & r_R : r^2 c \rightarrow rc.
\end{array}
$$

We identify the elements 1, r, r^2, c, rc, $r^2 c$ with the basis vectors $e_1, \ldots, e_6$, respectively, and translate the above as

$$
\begin{array}{lll}
r_R : e_1 \rightarrow e_2, & r_R : e_2 \rightarrow e_3, & r_R : e_3 \rightarrow e_1, \\
r_R : e_4 \rightarrow e_6, & r_R : e_5 \rightarrow e_4, & r_R : e_6 \rightarrow e_5.
\end{array}
$$

Hence the right regular representation for D_3 will require that

$$T(r) = \begin{bmatrix} 0 & 1 & 0 & 0 & 0 & 0 \\ 0 & 0 & 1 & 0 & 0 & 0 \\ 1 & 0 & 0 & 0 & 0 & 0 \\ 0 & 0 & 0 & 0 & 0 & 1 \\ 0 & 0 & 0 & 1 & 0 & 0 \\ 0 & 0 & 0 & 0 & 1 & 0 \end{bmatrix}.$$

Since c_R may be summarized as

$$\begin{array}{ccc} 1 \to c, & r \to rc, & r^2 \to r^2c, \\ c \to 1, & rc \to r, & r^2c \to r^2, \end{array}$$

the right regular representation for D_3 has

$$T(c) = \begin{bmatrix} 0 & 0 & 0 & 1 & 0 & 0 \\ 0 & 0 & 0 & 0 & 1 & 0 \\ 0 & 0 & 0 & 0 & 0 & 1 \\ 1 & 0 & 0 & 0 & 0 & 0 \\ 0 & 1 & 0 & 0 & 0 & 0 \\ 0 & 0 & 1 & 0 & 0 & 0 \end{bmatrix}.$$

The example can be concluded by repeating the argument above for r^2, rc, and r^2c, and by verifying that the required relations hold:

$$T(r)^3 = I_3, \qquad T(c)^2 = I_3, \qquad T(c)T(r) = T(r)^2T(c).$$

Now we are ready to state the formal definition.

11.4   Definition: Right regular representation of a finite group

Let $G = \{x_1, \ldots, x_n\}$ be a finite group of order n. For each x_i ($1 \le i \le n$), define a linear transformation $(x_i)_R$ on $\mathbf{R}^n$ by $(x_i)_R : e_j \to e_k$ whenever $x_j x_i = x_k$, the e_j being the natural basis for $\mathbf{R}^n$. Then $T(x_i)$ is the matrix associated with the linear transformation $(x_i)_R$.

It is helpful to note that every finite group has a right regular representation, but the form our discussion has taken is just a bit cumbersome.

Hence it is good to discover that the regular right representation can be constructed more easily from a suitably arranged group table.

Again, let $G = \{x_1, \ldots, x_n\}$ be a finite group of order n, and for each i, let $T(x_i)$ be the matrix associated with the linear transformation $(x_i)_R$. Since $T(x_i)$ merely permutes the vectors of the natural basis for $\mathbf{R}^n$, each matrix $T(x_i)$ will contain exactly one 1 in each row, one 1 in each column, and zeros in the remaining $n^2 - n$ positions. We'll give a name to such a matrix, motivated by our discussion of the groups S_n and A_n.

11.5   Definition: Permutation matrix

An $n \times n$ matrix having one 1 in each row, one 1 in each column, and zeros in the remaining $n^2 - n$ positions is called a *permutation matrix*.

Continuing our discussion of the right regular representation of a finite group G, notice that

$$
\begin{aligned}
T(x_i) \quad &\text{has a 1 as its } (j,k)\text{-entry} \\
&\text{if and only if } (x_i)_R : e_j \to e_k \\
&\text{if and only if } x_j x_i = x_k \\
&\text{if and only if } x_i = x_j^{-1} x_k,
\end{aligned}
$$

as pointed out earlier in this section. This tells us that the table for G must read, in part,

$$
\begin{array}{c|ccccc}
 & \cdot & \cdot & \cdot & x_k & \cdot \cdot \cdot \\
\hline
\vdots & & & & \vdots & \\
x_j^{-1} & & \cdots & & x_i & \\
\vdots & & & & &
\end{array}
$$

Thus if we prepare a group table whose *columns* are designated as $x_1, x_2, \ldots, x_n$ and whose *rows* are marked $x_1^{-1}, x_2^{-1}, \ldots, x_n^{-1}$, then $T(x_i)$ has a 1 in its (j,k)-entry if and only if the table described has x_i in row j, column k. Returning to Example 11.2, we may write the table for this cyclic group of order 3 as

$$
\begin{array}{c|ccc}
 & 1 & g & g^2 \\
\hline
1 & 1 & g & g^2 \\
g^2 & g^2 & 1 & g \\
g & g & g^2 & 1
\end{array}
$$

and read off $T(g)$ as the matrix having 1 where the table has g, and 0 elsewhere. Notice that the pattern of entries g^2 also matches the entries 1 in $T(g^2)$. In addition, the necessity for using $x_1^{-1}, x_2^{-1}, \ldots, x_n^{-1}$ for the rows is apparent from the fact that I_n must result as $T(1)$ from taking 1 as each diagonal product: hence, an entry on the main diagonal must represent $(x_j^{-1})(x_j)$.

Of course, there is no single order which must be followed in listing the elements of G as $x_1, x_2, \ldots, x_n$, although for convenience we'll always take $x_1 = 1$. A change in the order of the elements in this list, however, merely reflects a permutation of the natural basis vectors $e_1, \ldots, e_n$, which is a change of basis. In view of the discussion in Chapter 10, we therefore have the following.

11.6 Proposition: Equivalence of right regular representations

Two right regular representations for a finite group G, obtained by different orderings of the list of elements of G, are equivalent.

Exercises

11.1. Find the right regular representation for Z_5.

11.2. The left regular representation may be defined for a finite group G by merely considering multiplication on the left as an action of G on its own elements. Find the left and right regular representations for V_4 by direct computation (that is, *not* by using a suitably arranged group table), and observe that the two are identical (as they should be since V_4 is abelian).

11.3. Verify that if ζ is a primitive cube root of 1, then

$$U(g) = \begin{bmatrix} \zeta & 0 & 0 \\ 0 & \zeta^2 & 0 \\ 0 & 0 & 1 \end{bmatrix}$$

gives a representation for the cyclic group $\langle g \rangle$ of order 3. Then find a nonsingular matrix C with complex entries such that $CT(g) = U(g)C$, where T is the right regular representation found in Example 11.2. Then, without performing any explicit matrix multiplications, show that $CT(g^2) = U(g^2)C$ and $CT(1) = U(1)C$. What can you conclude about T and U?

11.4. Prove Proposition 11.6.

11.5. Consider a family of matrices of the form

$$\begin{bmatrix} \mathbf{0} & A \\ B & \mathbf{0} \end{bmatrix},$$

where A and B are $n \times n$ submatrices (note that they have the same degree), and $\mathbf{0}$ denotes an $n \times n$ submatrix of zeros. Show that the product of any two such matrices has the form:

$$\begin{bmatrix} C & \mathbf{0} \\ \mathbf{0} & D \end{bmatrix},$$

where C and D are $n \times n$ submatrices.

11.6. For the right regular representation T of D_3, find the remaining four matrices for T, and verify that T preserves the defining relations for D_3. Use Exercises 10.1 and 11.5 to simplify the computations.

11.7. Prove that the equivalence of representations, as defined in Definition 10.3, is an equivalence relation.

11.8. Let $G = \langle g : g^4 = 1 \rangle$ and form right regular representations T, U, and V by taking the elements of G in the orders

$$\{1, g, g^2, g^3\}, \qquad \{1, g, g^3, g^2\}, \qquad \text{and} \qquad \{1, g^3, g, g^2\},$$

respectively. Prove that T, U, and V are equivalent. (Note that Exercise 11.7 applies here.)

Chapter 12

Irreducible Representations

In Chapter 10 we found a representation V of degree 3 for D_4 such that *every* matrix $V(g)$ for $g \in D_4$ had the form

$$V(g) = \left[\begin{array}{cc|c} & & 0 \\ & T(g) & \\ & & 0 \\ \hline 0 & 0 & U(g) \end{array} \right]$$

where T and U were representations of D_4 of degree 2 and 1, respectively. Now consider the vector subspaces

$$S_1 = \{(x, y, 0) : x, y \in \mathbf{R}\},$$
$$S_2 = \{(0, 0, z) : z \in \mathbf{R}\}$$

of $\mathbf{R}^3$, which are simply the xy-plane and the z-axis. From linear algebra you know that $\mathbf{R}^3$ is the direct sum of S_1 and S_2, that is, that every vector in $\mathbf{R}^3$ can be written in exactly one way as a sum $s_1 + s_2$ with $s_1 \in S_1$ and $s_2 \in S_2$. To denote the direct sum we write

$$\mathbf{R}^3 = S_1 \oplus S_2.$$

If $v \in S_1$, then $vV(g) \in S_1$ also; likewise, if $w \in S_2$, then $wV(g) \in S_2$ as well, for any $g \in D_4$. In other words, all of the $T(g)$ leave the subspaces S_1 and S_2 invariant *as sets*, although the individual vectors in S_1 and S_2 are not necessarily left fixed.

12.1   Definition: Invariant subspace

Let T be a representation of a group G afforded by $\mathbf{R}^n$ for some n. If V is a vector subspace of $\mathbf{R}^n$ such that $v\,T(g) \in V$ for every $v \in V$ and every $g \in G$, then V is said to be *invariant under T*. The same applies if T is afforded by $\mathbf{C}^n$.

Although the term *invariant under T* does not specify the group G being considered, it is assumed that we know what that group is since T depends upon it.

With regard to the block form of $V(g)$ above, we noted informally in Chapter 10 that T and U are both representations of G. We'll now make this concept explicit and relate it to the idea of invariant subspaces.

12.2   Definition: Decomposable representation

Let T be a representation of G. We call T *decomposable* if there are representations A and B of G such that T is equivalent to the representation:

$$T^*(g) = \begin{bmatrix} A(g) & \mathbf{0} \\ \mathbf{0} & B(g) \end{bmatrix}, \qquad g \in G,$$

where the $\mathbf{0}$ terms denote rectangular arrays of zeros of the required size for T^* to be square. If T is not equivalent to any such representation, then T is called *indecomposable*.

Conversely, if T is equivalent to a representation of the form

$$\begin{bmatrix} A(g) & \mathbf{0} \\ \mathbf{0} & B(g) \end{bmatrix},$$

the fact that A and B are representations of G follows easily from Exercise 10.1 (see Exercise 12.1). Thus to show that a given T is decomposable, we need to prove only that it is equivalent to a representation T^* having the above form for all $g \in G$.

Bear in mind that if a representation T already has the form shown in Definition 12.2, then T is known to be decomposable; however, to show that a given T is indecomposable, we must show that T *is not equivalent to any* T^* having the form shown in Definition 12.2. Thus, for example, we do not yet know that the representation T found in Chapter 10 for the group of rigid symmetries of the cube is indecomposable because we have not shown that T is not equivalent to some T^* having the form of Definition 12.2.

As in Proposition 10.4, we should note that in order to show that a representation T is decomposable, it is sufficient to show that the form in Definition 12.2 is taken by all $T(x_j)$ and all $T(x_i^{-1})$, where G is generated by $x_1, \ldots, x_n$. In the case of finite G, the checks on the inverses may be omitted.

Let's now investigate the connection between decomposable representations and invariant subspaces. Suppose in Definition 12.1 that A has degree n and that B has degree m. Then T has degree $n + m$. Let $\{e_1, e_2, \ldots, e_{n+m}\}$ be the natural basis for $\mathbf{C}^{n+m}$ and let S_1 be the subspace spanned by $\{e_1, \ldots, e_n\}$ and S_2 be the subspace spanned by $\{e_{n+1}, \ldots, e_{n+m}\}$. Then $\mathbf{C}^{n+m} = S_1 \oplus S_2$ and each $T(g)$ carries vectors of S_1 to vectors of S_1 and vectors of S_2 to vectors of S_2. This claim should be evident from the form of the row vectors and of the matrices involved. We combine the ideas in the following items.

12.3   Definition: G-subspace

Let T be a representation of G having degree n and let S be a subspace of $\mathbf{C}^n$ with $1 \le \dim(S) < n$. If $T(g)$ carries vectors of S to vectors of S for all $g \in G$, then S is called a G-*subspace* of $\mathbf{C}^n$.

In this definition we have required $1 \le \dim(S) < n$ because $\dim(S) = n$ would mean that $S = \mathbf{C}^n$, and $\dim(S) = 0$ would mean $S = \{\mathbf{0}\}$, where $\mathbf{0}$ is the zero vector of $\mathbf{C}^n$; both of these are trivially G-subspaces of $\mathbf{C}^n$. Although we have written $\mathbf{C}^n$, we might just as well have used $\mathbf{R}^n$ if the representation T happened to be afforded by $\mathbf{R}$; the same alternative holds in what follows.

The terminology of this definition is standard but in a way unfortunate; it is clear that whether or not a subspace S is a G-*subspace* depends upon the particular representation T used as well as upon the group G, but T does not appear as part of the term defined. In practice, this hiatus will cause us no difficulty since the context will always make clear which representation T is under consideration.

12.4   Theorem: Criterion for decomposable represen-
tations

A representation T of a group G is decomposable if and only if the space $\mathbf{C}^n$ affording T has G-subspaces S_1 and S_2 such that $\mathbf{C}^n = S_1 \oplus S_2$.

Proof: If T is decomposable, then there is a change of basis that brings T into the form

$$T^*(g) = \begin{bmatrix} A(g) & \mathbf{0} \\ \mathbf{0} & B(g) \end{bmatrix}$$

for all $g \in G$. Then S_1 and S_2 can be formed as in the paragraph immediately preceding Definition 12.3 in terms of the new basis for $\mathbf{C}^n$. Conversely, if G-subspaces S_1 and S_2 exist with $\mathbf{C}^n = S_1 \oplus S_2$, then by a suitable change of basis (forming a new basis for $\mathbf{C}^n$ as the union of bases for S_1 and S_2), T can be brought into the form of T^* above. Specifically, if $\{x_1, \ldots, x_k\}$ and $\{x_{k+1}, \ldots, x_n\}$ are bases for S_1 and S_2, respectively, then the matrix C whose rows are the vectors $x_1, \ldots, x_k, x_{k+1}, \ldots, x_n$ gives the change of basis from the natural basis to the basis $\{x_1, \ldots, x_n\}$ for $\mathbf{C}^n$, and by hypothesis

$$CT^*(g) = T(g)C \qquad \text{for all } g \in G,$$

where T^* has the required form. This completes the proof.   □

It is important to realize that if $\mathbf{C}^n$ has a G-subspace S_1 for a representation T of G, then $\mathbf{C}^n$ will always have a *vector subspace* S_2 such that $\mathbf{C}^n = S_1 \oplus S_2$ as a direct sum of vector spaces, but S_2 *may fail to be a G-subspace of* $\mathbf{C}^n$. For example, consider the right regular representation of $V_4 = \{1, x, y, xy\}$ given by

$$T(x) = \begin{bmatrix} 0 & 1 & 0 & 0 \\ 1 & 0 & 0 & 0 \\ 0 & 0 & 0 & 1 \\ 0 & 0 & 1 & 0 \end{bmatrix}, \qquad T(y) = \begin{bmatrix} 0 & 0 & 1 & 0 \\ 0 & 0 & 0 & 1 \\ 1 & 0 & 0 & 0 \\ 0 & 1 & 0 & 0 \end{bmatrix}.$$

Now let $S_1 = \{(a, a, a, a) : a \in \mathbf{C}\}$; then $T(x)$ and $T(y)$ merely carry each point of S_1 to itself, and since S_1 is obviously a subspace of $\mathbf{C}^4$ of dimension 1, it is a V_4-subspace. If we arbitrarily choose $S_2 = \{(0, b, c, d) : b, c, d \in \mathbf{C}\}$, then $\mathbf{C}^4 = S_1 \oplus S_2$, but S_2 is *not* a V_4-subspace — observe that $(0, b, c, d)T(x) = (b, o, d, c) \notin S_2$ when $b \neq 0$. The difficulty here is that we took an obvious choice for S_2 to make $\mathbf{C}^4 = S_1 \oplus S_2$, but this is by no means the only possible choice for S_2, and the fact is that we can make a more fortuitous choice (see Exercise 12.8).

However, this example does suggest that it might be possible to have a G-subspace S_1 for which *no* complementary G-subspace S_2 exists. That would not detract from the fact that S_1 itself was a perfectly legitimate G-subspace. We'll consider this possibility in the next definition.

12.5   Definition: Reducible representation

Let T be a representation of a group G. We say that T is *reducible* if there exist representations A and B of G such that T is equivalent to the representation:

$$T^*(g) = \begin{bmatrix} A(g) & Q(g) \\ \mathbf{0} & B(g) \end{bmatrix},$$

where $\mathbf{0}$ is a submatrix of zeros and $Q(g)$ is a matrix (in general, not square) depending upon g. If T is not equivalent to such a representation, then T is *irreducible*.

In this definition $Q(g)$ has as many rows as $A(g)$ and as many columns as $B(g)$. Even if $Q(g)$ happens to be a square, it is never a representation of G because the upper right-hand corner of a product $T^*(g)T^*(h)$ would be $A(g)Q(h) + Q(g)B(h)$, spoiling the homomorphism.

The irreducible representations of a group G are the building blocks of all representations of G and will be a main focus of our study.

In view of the discussion in the Proof of Theorem 12.4, you should be able to prove the following theorem.

12.6   Theorem: Criterion for reducible representations

A representation T of a group G is reducible if and only if the space $\mathbf{C}^n$ affording T has a nontrivial G-subspace.

In Theorem 12.6 we assume, as in Definition 12.3, that $1 \le \dim(S) < n$, where S is the G-subspace in question and $n = \deg(T)$.

Naturally, we would not have made the discussion between *indecomposable* and *irreducible* if the two ideas were in fact equivalent, so let's consider an example of a representation that is indecomposable but reducible.

12.7   Example: An indecomposable representation that is reducible

Let $G = \langle x \rangle$ be an infinite cyclic group and consider the representation given by

$$T(x) = \begin{bmatrix} 1 & 1 \\ 0 & 1 \end{bmatrix}.$$

That T is a faithful representation of G with

$$T(x^n) = \begin{bmatrix} 1 & n \\ 0 & 1 \end{bmatrix}$$

for every integer n was shown in Chapter 9. Let $S_1 = \{(0, b) : b \in \mathbf{C}\}$; then for any vector in S_1 and any integer n, we have $(0, b)T(x^n) = (0, b)$, so S_1 is a G-subspace of $\mathbf{C}^2$ and T is reducible. Now if T is decomposable, then the G-subspaces we seek must both have dimension 1; hence, there must exist scalars λ and μ, and a nonsingular matrix

$$A = \begin{bmatrix} a & b \\ c & d \end{bmatrix}$$

such that

$$\begin{bmatrix} a & b \\ c & d \end{bmatrix} \begin{bmatrix} 1 & 1 \\ 0 & 1 \end{bmatrix} = \begin{bmatrix} \lambda & 0 \\ 0 & \mu \end{bmatrix} \begin{bmatrix} a & b \\ c & d \end{bmatrix},$$

that is,

$$\begin{bmatrix} a & a+b \\ c & c+d \end{bmatrix} = \begin{bmatrix} \lambda a & \lambda b \\ \mu c & \mu d \end{bmatrix}$$

and $\det(A) = ad - bc \neq 0$. If such scalars and A exist, we'll have shown that T is equivalent to a representation given by

$$T^*(x) = \begin{bmatrix} \lambda & 0 \\ 0 & \mu \end{bmatrix}.$$

Equating entries above, we have $\lambda a = a$, so either $\lambda = 1$ or $a = 0$. If $\lambda = 1$, then $a + b = \lambda b = b$, so $a = 0$. On the other hand, if $a = 0$, then $ad - bc \neq 0$ gives $bc \neq 0$, which means that $b \neq 0$ and $c \neq 0$. But we also have $\lambda b = a + b = b$ (using $a = 0$), so $\lambda = 1$ (since $b \neq 0$). Thus, *both* $\lambda = 1$ *and* $a = 0$ hold (since each of these conditions implies the other, and at least one must hold). We thus have

$$\begin{bmatrix} 0 & b \\ c & c+d \end{bmatrix} = \begin{bmatrix} 0 & b \\ \mu c & \mu d \end{bmatrix}$$

together with $bc \neq 0$. Again equating entries, we see that $\mu c = c$ yields $\mu = 1$ (since $c \neq 0$). Hence $c + d = \mu d = d$ gives $c = 0$, which contradicts $bc \neq 0$. Therefore, no such λ, μ, and A can exist, and T is indecomposable.

Directly from the relevant definitions, we have the following.

12.8   Proposition: Decomposable implies reducible

If a representation T of a group G is decomposable, then T is reducible. Equivalently, if T is irreducible, then T is indecomposable.

Example 12.7 shows that a representation may indeed be reducible and indecomposable. The use of an infinite group for this example is not coincidental; in fact, the situation for finite groups is much happier, as was shown by Maschke in 1898. To formulate Maschke's result, we need another basic definition.

12.9   Definition: Completely reducible representation

A representation T of a group G is *completely reducible* if whenever T has a G-subspace S_1, T also has a G-subspace S_2 such that $\mathbf{C}^n = S_1 \oplus S_2$, where n is the degree of T.

In terms of matrices, a representation T is completely reducible if whenever it can be put (by means of equivalence) into the form shown in Definition 12.5, it can also be put into the form shown in Definition 12.2. But bear in mind that putting T into either of these forms means finding *one single* change-of-basis matrix that puts *all* $T(g)$ into the proper form.

From Definitions 12.5 and 12.9 follows a remark that is linguistically amusing, though mathematically trivial:

Every irreducible representation is completely reducible.

This is of no consequence on Earth, but I hope it gave you a chuckle. (On the other hand, you should be able to explain *why* it is a true statement!)

Now we are ready to consider the statement of Maschke's theorem.

12.10   Theorem: Maschke's theorem

If G is a finite group and T is a representation of G afforded by $\mathbf{C}^n$ (or by $\mathbf{R}^n$), then T is completely reducible.

We won't prove this theorem here, but we'll make extensive use of the result. I should note, for those familiar with the terminology, that the above is actually a special case of Maschke's theorem. In fact it holds not only for $\mathbf{C}$ and $\mathbf{R}$, but for any field K whose characteristic p does not divide the order of the group G.

To investigate the implications of Maschke's theorem, we'll let T be a reducible representation of a finite group G. Then T is equivalent to a representation T^* of the form in Definition 12.5. But Maschke's theorem then tells us that T is equivalent to a representation T^{**} of the form in Definition 12.2. Let's write

$$T^{**}(g) = \begin{bmatrix} A(g) & \mathbf{0} \\ \mathbf{0} & B(g) \end{bmatrix}, \qquad \text{for all } g \in G.$$

Now Maschke's theorem tells us that A and B are also completely reducible. Thus either A is irreducible, or we have

$$A(g) = \begin{bmatrix} A_1(g) & \mathbf{0} \\ \mathbf{0} & A_2(g) \end{bmatrix}, \qquad \text{for all } g \in G.$$

Since the degree of A is finite, we can repeat this decomposition at most a finite number of times until we have a string of irreducible representations $A_1(g), \ldots, A_k(g)$ down the diagonal of the matrix. Of course, each time we make such a decomposition, we are finding an equivalent representation, so we are exploiting the idea that equivalence is transitive (see Exercise 11.7). Then the same process of decomposition may be applied to the B in T^{**}, and we may denote these irreducible blocks by $A_{k+1}(g), \ldots, A_n(g)$. The result is that our original representation T is equivalent to a representation of the form

$$T^{***}(g) = \begin{bmatrix} A_1(g) & \mathbf{0} & \cdots & \mathbf{0} \\ \mathbf{0} & A_2(g) & \cdots & \mathbf{0} \\ \cdots & \cdots & \cdots & \cdots \\ \mathbf{0} & \mathbf{0} & \cdots & A_n(g) \end{bmatrix}, \qquad \text{for all } g \in G,$$

in which all of $A_1(g), \ldots, A_n(g)$ are irreducible. We've thus proved the following theorm.

12.11 Theorem: Decomposition into irreducible components

Every representation T of a finite group G over **R** or **C** may be decomposed into irreducible representations $A_1, \ldots, A_n$ in the sense that T is equivalent to a representation T^* in which the $A_1, \ldots, A_n$ appear down the diagonal of T^*, and the rest of each $T^*(g)$ consists of blocks of zeros. We call the $A_1, \ldots, A_n$ the *irreducible components* of T.

12.12 Example: Decomposition of the regular representation of V_4

The right regular representation of $V_4 = \{1, x, y, xy\}$ is given by

$$T(x) = \begin{bmatrix} 0 & 1 & 0 & 0 \\ 1 & 0 & 0 & 0 \\ 0 & 0 & 0 & 1 \\ 0 & 0 & 1 & 0 \end{bmatrix}, \qquad T(y) = \begin{bmatrix} 0 & 0 & 1 & 0 \\ 0 & 0 & 0 & 1 \\ 1 & 0 & 0 & 0 \\ 0 & 1 & 0 & 0 \end{bmatrix}.$$

A one-dimensional representation is necessarily irreducible, and (see Exercise 10.2) the one-dimensional representations of V_4 are given by

$$\begin{array}{ll} T_1(x) = [1], & T_1(y) = [1]; \\ T_2(x) = [1], & T_2(y) = [-1]; \\ T_3(x) = [-1], & T_3(y) = [1]; \\ T_4(x) = [-1], & T_4(y) = [-1]. \end{array}$$

Now let

$$C = \begin{bmatrix} 1 & 1 & 1 & 1 \\ 1 & 1 & -1 & -1 \\ -1 & 1 & -1 & 1 \\ -1 & 1 & 1 & -1 \end{bmatrix};$$

then $\det(C) = -16$, so C is nonsingular. By performing the matrix products you can verify that

$$CT(x) = \begin{bmatrix} 1 & 0 & 0 & 0 \\ 0 & 1 & 0 & 0 \\ 0 & 0 & -1 & 0 \\ 0 & 0 & 0 & -1 \end{bmatrix} C \quad \text{and} \quad CT(y) = \begin{bmatrix} 1 & 0 & 0 & 0 \\ 0 & -1 & 0 & 0 \\ 0 & 0 & 1 & 0 \\ 0 & 0 & 0 & -1 \end{bmatrix} C.$$

Hence the right regular representation of V_4 is equivalent to a representation whose irreducible components are precisely the four one-dimensional representations of V_4.

As remarked in the preceding example, a one-dimensional representation is obviously irreducible, so it will be helpful to show that not all irreducible representations must have degree 1. The following example will serve.

12.13 Example: An irreducible representation of degree 3

In Chapter 10 we found a representation of the group G of rigid symmetries of the cube given by

$$T(r) = \begin{bmatrix} 0 & -1 & 0 \\ 1 & 0 & 0 \\ 0 & 0 & 1 \end{bmatrix}, \quad T(d) = \begin{bmatrix} 0 & -1 & 0 \\ 0 & 0 & -1 \\ 1 & 0 & 0 \end{bmatrix}.$$

Now, by Maschke's theorem, if T is reducible, it must be equivalent to a representation having the form

$$T^*(g) = \left[\begin{array}{cc|c} & & 0 \\ & U(g) & 0 \\ \hline 0 & 0 & V(g) \end{array} \right].$$

Here we have arbitrarily chosen a 2×2 in the upper left and 1×1 in the lower right; the reverse arrangement would be equivalent, as you are asked to check in Exercise 12.3. If such an equivalence of T to T^* exists, we must have a *nonsingular* C for which $CT(g) = T^*(g)C$ for all $g \in G$. Now suppose that $V(r) = [\lambda]$ and $V(d) = [\mu]$, where λ and μ are complex numbers. From the *third row* of the products

$$C \begin{bmatrix} 0 & -1 & 0 \\ 1 & 0 & 0 \\ 0 & 0 & 1 \end{bmatrix} = \left[\begin{array}{cc|c} & & 0 \\ & U(r) & 0 \\ \hline 0 & 0 & \lambda \end{array} \right] C$$

we obtain

$$c_{32} = \lambda c_{31}, \qquad -c_{31} = \lambda c_{32}, \qquad c_{33} = \lambda c_{33},$$

and from the *third row* of the products

$$C \begin{bmatrix} 0 & -1 & 0 \\ 0 & 0 & -1 \\ 1 & 0 & 0 \end{bmatrix} = \left[\begin{array}{cc|c} & & 0 \\ & U(d) & 0 \\ \hline 0 & 0 & \mu \end{array} \right] C$$

we obtain

$$c_{33} = \mu c_{31}, \qquad -c_{31} = \mu c_{32}, \qquad -c_{32} = \mu c_{33}.$$

From the third of these six equations involving λ and μ we conclude that $\lambda = 1$ or $c_{33} = 0$. But if $c_{33} = 0$, then (from the last three equations) $c_{31} = c_{32} = 0$, which gives C an entire row of zeros and makes C singular, contrary to our assumption. Hence $\lambda = 1$. But then the first two equations give $c_{31} = c_{32} = c_{33} = 0$, making C singular. Since both attempts lead to a contradiction, no such C can exist, and T must be irreducible.

You'll note that in Example 12.13 we made no assumption whether the entries c_{ij} in C were real or complex; in neither case would T be equivalent to the postulated T^*. To conclude this section, let's look at an example of a representation that is irreducible over **R** but reducible over **C**.

12.14 Example: A representation reducible over C but not over R

Let $G = \langle g \rangle$ be cyclic of order 3, and take

$$U(g) = \begin{bmatrix} 0 & -1 \\ 1 & -1 \end{bmatrix};$$

then $U(g)$ determines a representation of G. If U is reducible over the real numbers $\mathbf{R}$, then there must exist a nonsingular matrix A and real scalars λ and μ with

$$AU(g) = \begin{bmatrix} \lambda & 0 \\ 0 & \mu \end{bmatrix} A.$$

Apply to this equation the fact that the determinant of a product equals the product of the determinants; this gives $\det(A) \cdot 1 = \lambda\mu \cdot \det(A)$. Since $\det(A) \neq 0$, it follows that $\lambda\mu = 1$. Now the *trace* of a square matrix is the sum of the elements on the main diagonal. In general, the trace of a product is *not* equal to the product of the traces; however, we'll see in Chapter 14 that if there is a nonsingular matrix C such that $B = C^{-1}AC$, then A and B have the same trace. Thus, the matrix equation above shows that $\text{trace}(U(g)) = \lambda + \mu$; that is, $-1 = \lambda + \mu$. But then,

$$\lambda\mu = \lambda(-1 - \lambda) = 1,$$

whence

$$\lambda^2 + \lambda + 1 = 0,$$

which is an equation having no real root. Therefore, U is irreducible over $\mathbf{R}$. The fact that U is reducible over $\mathbf{C}$ is left as Exercise 12.4.

Exercises

12.1. Verify that if a representation T of a group G is equivalent to a representation of the form

$$T^*(g) = \begin{bmatrix} A(g) & 0 \\ 0 & B(g) \end{bmatrix},$$

then A and B are representations of G.

12.2. Prove Theorem 12.6.

12.3. Let A and B be representations of G, and for each $g \in G$ let

$$T(g) = \begin{bmatrix} A(g) & 0 \\ 0 & B(g) \end{bmatrix} \quad \text{and} \quad T^*(g) = \begin{bmatrix} B(g) & 0 \\ 0 & A(g) \end{bmatrix}.$$

Prove that T is equivalent to T^*.

12.4. Verify that the representation U in Example 12.14 is reducible over **C** by finding a nonsingular matrix A and complex scalars λ and μ such that the equation for equivalence of representations holds.

12.5. Let ζ be a primitive kth root of 1, let $G = \langle x : x^{2k} = 1 \rangle$, and let U be a representation of G given by

$$U(x) = \begin{bmatrix} 0 & \zeta \\ 1 & 0 \end{bmatrix}.$$

Verify that U is indeed a representation of G, and find its irreducible components. Then check that U is *not* a representation of a cyclic group of order k. (Caution: k need not be even.)

12.6. For the representation T of $V_4 = \{1, x, y, xy\}$ given by

$$T(x) = \begin{bmatrix} 0 & 1 \\ 1 & 0 \end{bmatrix}, \qquad T(y) = \begin{bmatrix} -1 & 0 \\ 0 & -1 \end{bmatrix},$$

find the irreducible components of T.

12.7. Use determinants and traces to give an alternative proof of the fact that the representation of the infinite cyclic group $\langle x \rangle$ given by

$$T(x) = \begin{bmatrix} 1 & 1 \\ 0 & 1 \end{bmatrix}$$

is indecomposable.

12.8. Consider the representation T in Example 12.12. Let S_1 be the subspace of $\mathbf{R}^4$ spanned by the vector $(1, 1, 1, 1)$, and let S_2 be the subspace spanned by the set

$$\{(1, 1, -1, -1), (1, -1, 1, -1), (1, -1, -1, 1)\}.$$

Show that $\mathbf{R}^4 = S_1 \oplus S_2$ and that S_1 and S_2 are V_4-subspaces of $\mathbf{R}^4$.

Chapter 13

Representations of Abelian Groups

We have already seen a number of representations of abelian groups of various orders (including the infinite cyclic group). In this chapter we'll find all *one-dimensional* representations of *finite* abelian groups and then apply that knowledge to finding one-dimensional representations of nonabelian finite groups.

First, let $G = \langle x \rangle$ be a cyclic group of order k, and ζ a primitive kth root of 1. Now to determine a representation T of G, we need only to give $T(x)$ and to verify that $(T(x))^k = I_n$, the $n \times n$ identity matrix, where n is the degree of T. In particular, if $T(x) = [\zeta^j]$, where $1 \leq j \leq k$, then $(T(x))^k = [1]$, as required. Conversely, if $T(x) = [\theta]$, where $\theta \in \mathbf{C}$, then $(T(x))^k = [1]$ requires that $\theta^k = 1$, so θ must be some power of ζ. This, along with Exercise 10.10, has proved the following.

13.1 Proposition: The one-dimensional representations of $\mathbf{Z}_k$

If G is cyclic of order k, and if ζ is a primitive kth root of 1, then for $j = 1, 2, \ldots, k$, $T_j(x) = [\zeta^j]$ is a complete list of the one-dimensional representations of G.

We'll need to consider some additional material from abstract group theory, but to motivate that discussion, you should think about the fact that V_4 consists of two cyclic groups, which we may denote as $\langle x \rangle$ and $\langle y \rangle$, together with the product xy that they determine. This is the prototype for the idea of the direct product.

13.2 Definition: Internal direct product

Let H and K be normal subgroups of G with $H \cap K = \{1\}$, and $G = HK$. Then G is called the *internal direct product* of H and K; we write $G = H \times K$.

It should be clear at this point that $V_4 = \langle x \rangle \times \langle y \rangle$ is an example of a direct product. The use of the term *internal* reflects the fact that we are viewing G as the direct product of two of its own subgroups; an opposite approach will appear shortly, but first we'll look at two useful characterizations of the internal direct product.

13.3 Proposition: Characterizations of the direct product

Let H and K be subgroups of G. Then the following statements are equivalent:

(i) $G = H \times K$;

(ii) $G = HK$, $H \cap K = \{1\}$, and $hk = kh$ for all $h \in H$ and $k \in K$;

(iii) H and K are normal subgroups of G, and each element of G can be written in exactly one way as a product hk with $h \in H$ and $k \in K$.

The proof of this proposition is left to the Exercises.

The discussion of groups of order 35 in Example 7.16 actually amounted to showing that such a group was the internal direct product of a subgroup of order 5 and a subgroup of order 7. Exercise 7.1 did a similar thing for a group of order 15. In both of these instances, the resulting group was cyclic, but the example of V_4 cited above shows that a direct product of cyclic groups need not necessarily be cyclic. A criterion for such a direct product to be cyclic is that the two subgroups H and K have relatively prime order. We'll prove this result later in this chapter.

The procedure of writing a group as an internal direct product of two of its subgroups can be reversed to take two given groups and to form a new group as their direct product. This is the most straightforward way (but not the only one) of forming larger groups from smaller ones.

13.4   Proposition: External direct product

Let H and K be groups, let $G = \{(h,k) : h \in H, k \in K\}$, define an operation in G by $(h,k)(h',k') = (hh', kk')$, let $H^* = \{(h,1) : h \in H\}$ and $K^* = \{(1,k) : k \in K\}$. Then G is a group, H^* and K^* are subgroups of G, $H \cong H^*$, $K \cong K^*$, and $G = H^* \times K^*$. Here G^* is called the *external direct product of H and K*.

Proof: Closure and associativity in G are immediate from the fact that H and K are groups. The identity of G is $(1,1)$, where we have used 1 to represent the identity of H and of K. Clearly, (h^{-1}, k^{-1}) serves as the inverse of the element (h,k). The two isomorphisms are immediate (just from the notation). To show that $G = H^* \times K^*$ we need to establish that H^* and K^* are normal in G. But for $(h,1) \in H^*$ and $(h',k') \in G$, we have

$$
\begin{aligned}
(h',k')^{-1}(h,1)(h',k') &= (h'^{-1}, k'^{-1})(h,1)(h',k')\\
&= (h'^{-1}hh', k'^{-1}k')\\
&= (h'^{-1}hh', 1) \in H^*,
\end{aligned}
$$

and a similar argument holds for $(1,k)$. Certainly, $G = H^*K^*$ and $H^* \cap K^* = \{(1,1)\}$. □

Note in the above that the external direct product G^* is the internal direct product of its subgroups H^* and K^*. Similarly, if G is an internal direct product of two of its subgroups, it is isomorphic to their external direct product. We'll make this explicit in the following.

13.5   Proposition: Internal and external direct products are isomorphic

Let $G = H \times K$ and define $G^* = \{(h,k) : h \in H, k \in K\}$ with the operation and the subgroups H^* and K^* as in Proposition 13.4. Then $H^* \cong H$, $K^* \cong K$, and $G^* \cong G$.

Thus the difference between internal and external direct products is merely the point of view: whether the group G is already given, or G is formed by putting together two given groups. Ordinarily, we may simply write hk in place of the notation (h,k) of Proposition 13.4.

Now suppose that $G = H \times K$, T is a representation of H and U is a representation of K. It would be convenient if we could take advantage of the decomposition of each $g \in G$ in a unique way as a product hk (as in Proposition 13.3iii) to define a representation V of G. However, from Proposition 13.3ii we know that $hk = kh$ for any choice of $h \in H$ and $k \in K$, which means that (in the notation of Proposition 13.4)

$$(h, 1)(1, k) = (h, k) = (1, k)(h, 1),$$

and thus to define a representation of G by $V(g) = T(h)U(k)$, we would need to have $T(h)U(k) = U(k)T(h)$ in all cases, which is not true of matrix products in general. Still, we've seen that one-dimensional representations are not unimportant, and 1×1 matrices do commute.

13.6   Proposition: One-dimensional representation of direct product

If $G = H \times K$ and if T and U are one-dimensional representations of H and K, respectively, then $V(hk) = T(h)U(k)$ is a representation of G.

For example, consider the group G of order 9 given by

$$G = \langle x, y : x^3 = y^3 = 1, yx = xy \rangle;$$

then $G = \langle x \rangle \times \langle y \rangle$. Both $\langle x \rangle$ and $\langle y \rangle$ are of order 3, and so the only one-dimensional representations available are (by Proposition 13.1) those determined by

$$
\begin{array}{lll}
T_1(x) = [1], & T_2(x) = [\zeta], & T_3(x) = [\zeta^2]; \\
U_1(y) = [1], & U_2(y) = [\zeta], & U_3(y) = [\zeta^2],
\end{array}
$$

where ζ is a primitive cube root of 1. This gives us nine representations of the form

$$V(x^r y^s) = T_j(x)^r U_k(y)^s$$

for $r, s = 1, 2, 3$ and $j, k = 1, 2, 3$ also. Could G have any other one-dimensional representations? If V is any one-dimensional representation of G, then the restriction of its domain to the subgroup $\langle x \rangle$ gives a representation of that subgroup; hence $V(x) = T_j(x)$ for some $j = 1, 2, 3$. Similarly, $V(y) = U_k(y)$ for some $k = 1, 2, 3$. Thus V must agree with some T_j on $\langle x \rangle$ and with some U_k on $\langle y \rangle$. But G consists simply of products of powers of x and y, so V is completely determined by the T_j and U_k in question;

that is, $V = T_j U_k$ for some choice of $j, k = 1, 2, 3$. Moreover, by Exercise 10.10, we know that these nine representations of G are inequivalent. We have found all of the one-dimensional representations of this group.

Happily, the argument used in the preceding example is the prototype for finding all of the one-dimensional representations of a finite abelian group. Note first that, of course, the process of forming an external direct product can be repeated as follows.

13.7 Definition: Direct product of n groups

Given groups $H_1, H_2, \ldots, H_n$, we can form the external direct product by defining

$$G = (h_1, h_2, \ldots, h_n), \qquad \text{where } h_j \in H_j \text{ for each } j$$

and taking the operation component by component, as in Proposition 13.4. Conversely, let $H_1, H_2, \ldots, H_n$ be normal subgroups of a group G and let each element $g \in G$ have a unique representation as a product $g = h_1 h_2 \cdots h_n$ with $h_j \in H_j$ for each j. Then G is called the *direct product of* $H_1, H_2, \ldots, H_n$; we write

$$G = H_1 \times H_2 \times \cdots \times H_n.$$

In this definition we took Proposition 13.3iii as the model rather than Definition 13.2 because it gives the simpler form when there are more than two factors. An analogue of Definition 13.2 can be shown to hold, but it must be stated carefully.

The next theorem will be stated without proof but (like Maschke's theorem in Theorem 12.10) used extensively.

13.8 Theorem: Main theorem on finite abelian groups

If G is a finite abelian group, then G may be written as a direct product of cyclic groups in either of the following ways:

(i) $G \cong Z_{q_1} \times \cdots \times Z_{q_n}$, where each q_j is a power of a prime;

(ii) $G \cong Z_{r_1} \times \cdots \times Z_{r_m}$, where r_j divides r_{j+1} for $1 \leq j \leq m - 1$.

Let's illustrate this result with some examples.

13.9   Examples: Some finite abelian groups

Let $|G| = 24$. Since $24 = 2^3 \cdot 3$, the possibilities for form (i) are

$$Z_8 \times Z_3,$$
$$Z_4 \times Z_2 \times Z_3,$$
$$\text{and} \quad Z_2 \times Z_2 \times Z_2 \times Z_3.$$

All three of these groups exist; they are isomorphic, respectively, to

$$Z_{24},$$
$$Z_2 \times Z_{12},$$
$$\text{and} \quad Z_2 \times Z_2 \times Z_6,$$

each of which has the form shown in (ii). By the theorem, there are no other abelian groups of order 24.

Let $|G| = 21$. Since $21 = 3 \cdot 7$, there is only one possibility for an abelian group of this order, namely, in form (i), $Z_3 \times Z_7$, which is isomorphic to Z_{21} in form (ii). We remarked at the end of Exercise 7.1 that there is a noncyclic group of order 21; you now know that this group must be nonabelian as well.

Let $|G| = 16$. Since $16 = 2^4$, the abelian groups of this order are given by

$$Z_{16},$$
$$Z_2 \times Z_8,$$
$$Z_4 \times Z_4,$$
$$Z_2 \times Z_2 \times Z_4,$$
$$\text{and} \quad Z_2 \times Z_2 \times Z_2 \times Z_2.$$

Note that these decompositions are valid for both forms, (i) and (ii).

The isomorphisms asserted to exist in the first two examples may be easily seen from the following result.

13.10 Proposition: $Z_{mn} = Z_m \times Z_n$ for relatively prime m and n

If m and n are relatively prime integers, then the cyclic group of order mn is isomorphic to $Z_m \times Z_n$.

Proof: Let G be the external direct product of $\langle a : a^m = 1 \rangle$ and $\langle b : b^n = 1 \rangle$; then G may be written as $\langle a, b : a^m = b^n = 1, ba = ab \rangle$. Let $H = \langle x : x^{mn} = 1 \rangle$, and define $\sigma : H \to G$ by $\sigma(x) = ab$; this means that, for $0 \le k < mn$, we take $\sigma(x^k) = a^k b^k$. Then $\sigma(x^i x^j) = \sigma(x^{i+j}) = a^{i+j} b^{i+j} = a^i b^i a^j b^j = \sigma(x^i)\sigma(x^j)$, so σ is a homomorphism. Let $x^k \in \ker(\sigma)$; then $a^k b^k = 1$, and hence both m and n must divide k. Since m and n are relatively prime and $0 \le k < mn$, we must have $k = 0$, that is, $x^k = 1$; therefore, σ is one-to-one. Now $|G| = |\langle a \rangle| \cdot |\langle b \rangle| = mn = |H|$, and σ is $1 : 1$, so σ is onto as well. □

Theorem 13.8 says that if G is an *abelian* group of order p^2, where p is a prime, then either G is cyclic of order p^2 or G is isomorphic to $Z_p \times Z_p$. In particular, the only abelian groups of order 4 are Z_4 and $Z_2 \times Z_2$, the latter being isomorphic to V_4. We can generalize this observation as follows.

13.11 Proposition: Groups of order p^2

Let p be a prime and $|G| = p^2$, then either G is cyclic or G is isomorphic to $Z_p \times Z_p$. Hence, G is abelian.

Proof: Observe first that the cyclic group of order p^2 is not isomorphic to $Z_p \times Z_p$ because the cylic group has elements of order p^2 and $Z_p \times Z_p$ does not. Clearly, both of these groups are abelian. Now suppose that G is not cyclic; we'll prove that $G \cong Z_p \times Z_p$. By Proposition 7.2, the center $C(G)$ contains a nonidentity element a. Now the order of a must be p. Since $a \in C(G)$, $g^{-1}ag = a$ for all $g \in G$, and so $A = \langle a \rangle$ is a normal subgroup of G. If $b \notin A$, then b has order p, and the distinct cosets of A in G are A, Ab, $Ab^2, \ldots, Ab^{p-1}$, accounting for all of the p^2 elements of G. Let $B = \langle b \rangle$; then $G = AB$, $A \cap B = \{1\}$, and since $ab = ba$, B is normal in G also. By Proposition 13.2, $G = A \times B$. This completes the proof. □

By means of Proposition 13.6 and Theorem 13.8 we can easily determine all one-dimensional representations of a finite abelian group G. Let's write G in form (i) of Theorem 13.8 and write $Z_{q_j} = \langle x_j \rangle$ for $j = 1, 2, \ldots, n$. If ζ_j is a primitive q_jth root of 1, then the list

$$T_{j1}(x_j) = [1], \quad T_{j2}(x_j) = [\zeta_j], \quad T_{j3}(x_j) = [\zeta_j^2], \quad \ldots, \quad T_{jq_j}(x_j) = [\zeta_j^{q_j-1}]$$

gives all one-dimensional representations of $\langle x_j \rangle$, and we may simply take products of these T_{jk} as we did in Proposition 13.6 to obtain all of the one-dimensional representations of G. There are $q_1 q_2 \cdots q_n$ of these, but this product is just $|G|$. Thus we have proved the following theorem.

13.12　Theorem: One-dimensional representations of a finite abelian group

If G is a finite abelian group, then there are exactly $|G|$ one-dimensional representations of G over the complex numbers $\mathbf{C}$.

We'll see later that the one-dimensional representations are the *only* irreducible representations of a finite abelian group over $\mathbf{C}$.

To conclude this section, we'll show how the one-dimensional representations of finite *abelian* groups may be used to determine the one-dimensional representations of an arbitrary finite group. The first step is to show how any representation of a factor group G/N gives rise to a representation of the group G.

13.13　Proposition: A representation of a group from one of a factor group

Let $N \trianglelefteq G$ and let T be any representation of the factor group G/N with degree n. Define T^* on G by $T^*(g) = T(Ng)$. Then T^* is a representation of G. Moreover, if U is a representation of G/N for which U is not equivalent to T, then U^* (formed as was T^*) is not equivalent to T^*.

Proof: First we show that T^* preserves products. Here

$$T^*(gh) = T(N(gh)) = T((Ng)(Nh)) = T(Ng)T(Nh) = T^*(g)T^*(h),$$

so T^* is indeed a representation of G. If U^* is equivalent to T^*, then there exists a nonsingular matrix A such that $AU^*(g) = T^*(g)A$ for all $g \in G$. But then it is also true that $AU(Ng) = T(Ng)A$ for all $Ng \in G/N$, so U is equivalent to T. □

For a simple example, let

$$G = D_4 = \langle r, c : r^4 = c^2 = 1, cr = r^{-1}c \rangle,$$

let $N = \langle r \rangle$, and consider the representation:

$$T(N) = [1], \qquad T(Nc) = [-1]$$

on the factor group G/N, which has order 2. Then,

$$T^*(r^j) = [1] \qquad \text{and} \qquad T^*(r^j c) = [-1]$$

for $j = 0, 1, 2, 3$.

To apply this construction of a representation of G from one of G/N, we need the concept of a derived group.

13.14   Definition: Derived group

The *derived group of G* (also called the *commutator subgroup of G*) is the subgroup G' generated by all elements of the form $ghg^{-1}h^{-1}$ for $g, h \in G$. Such an element is called a *commutator* and is sometimes denoted as $[g, h]$.

It's important to realize that closure in the derived group G' is guaranteed by the phrase "generated by." Moreover, bear in mind that the elements G' are *products* of commutators; they need not all be commutators themselves.

In a sense, the commutator $[g, h]$ tells "how far the elements g and h are from commuting with one another." Since

$$gh = (ghg^{-1}h^{-1})(hg) = [g, h](hg),$$

we might say that gh and hg are "within $[g, h]$ of commuting." For example, you can easily check that $D_4' = \{1, r^2\}$, so elements of D_4 either commute or are within r^2 of commuting.

13.15   Proposition: Abelian factor groups and the derived group

For any group G, the derived group G' is a normal subgroup. Moreover, if $N \trianglelefteq G$, then G/N is abelian if and only if $G' \leq N$.

A special case of Proposition 13.15 is the one in which $N = G'$; along with Propositions 13.12 and 13.13. This establishes the fact that there are *at least* $[G : G']$ one-dimensional representations of a finite group G (or even of an infinite group G with the property that G/G' is finite). We'll shortly see that "at least" may be replaced by "exactly."

13.16   Proposition: One-dimensional representations of G and G/G'

Let T be any one-dimensional representation of a group G. Then T, with its domain restricted to G', is the trivial representation of G'. Moreover, if we define $T^{\#}(G'x) = T(x)$ for each $G'x \in G/G'$, then $T^{\#}$ is a representation of G/G'.

Finally, we reach our main result, the following theorem.

13.17   Theorem: One-dimensional representations of a finite group

If G is a finite group, then G has exactly $[G : G']$ one-dimensional representations over the complex numbers $\mathbf{C}$.

Proof: By the discussion following Proposition 13.15, there are *at least* $[G : G']$ one-dimensional representations of G over the complex numbers. By Proposition 13.16, every such representation gives rise to one of G/G'; hence, there can be no more than $[G : G']$ of them.   □

Theorem 13.17 holds under the weaker hypothesis that G/G' is a finite group, even if G itself is infinite.

Exercises

13.1. By considering $hkh^{-1}k^{-1}$, show that (i) implies (ii) in Proposition 13.3. Then show that if you assume (ii), you can prove normality, which is all you need to show that (ii) implies (i).

13.2. To show that (i) implies (iii) in Proposition 13.3, show that if $h_1 k_1 = h_2 k_2$ with $h_1, h_2 \in H$ and $k_1, k_2 \in K$, then $h_1 = h_2$ and $k_1 = k_2$. Then show that (iii) implies (i).

13.3. Find all abelian groups of orders 18, 32, and $\overset{72}{36}$. (Give both forms from Theorem 13.8.)

13.4. To illustrate Proposition 13.10 and Theorem 13.12, let $p = 2$ and $q = 3$. Take $\zeta = -1$ and $\eta = \frac{1}{2}(-1 + i\sqrt{3})$, a primitive cube root of 1. Verify that $\zeta\eta$ (of course, $\zeta\eta = -\eta$) is a primitive sixth root of 1 and that the one-dimensional representations of Z_6 you get from the proof of Theorem 13.12 are the same as those you get from Proposition 13.1.

13.5. Show that $D_4 = \langle r, c \rangle$ is *not* the direct product of $\langle r \rangle$ and $\langle c \rangle$.

13.6. The dihedral groups D_3 and D_6 are, respectively, the groups of symmetries of the equilateral triangle and the regular hexagon. Let $G = D_3 \times Z_2$, written as

$$G = \langle x, y : x^3 = y^2 = 1, yx = x^{-1}y \rangle \times \langle z : z^2 = 1 \rangle$$

and write $D_6 = \langle r, c : r^6 = c^2 = 1, cr = r^{-1}c \rangle$. Find an isomorphism ϕ from D_6 onto G.

13.7. Prove Proposition 13.15.

13.8. Show that the derived group $G' = \{1\}$ if and only if G is abelian.

13.9. Show that the derived group A_5' is equal to A_5 itself.

13.10. Use Theorem 13.17 to find all one-dimensional representations of Q_2.

13.11. Use Theorem 13.17 to find all one-dimensional representations of D_4.

13.12. Use Theorem 13.17 to find all one-dimensional representations of A_4.

13.13.   Prove Proposition 13.16. Note that it is necessary to show that $T^\#$
         is well defined (recall Proposition 6.5); specifically, if $G'x = G'y$,
         then $T^\#(G'x) = T^\#(G'y)$.

Chapter 14

Group Characters

We'll now turn to an important function on a group, one that arises in many applications.

14.1   Definition: Group character

The *trace* of a square matrix A is the sum of the elements on the main diagonal; we'll denote the trace of A as $tr(A)$. If A is a matrix of degree n written as (a_{ij}), then

$$tr(A) = \sum_{i=1}^{n} a_{ii}.$$

Now let T be a representation of a group G, and for each $g \in G$, let $\chi(g) = tr(T(g))$. We call χ a *character* of G, in particular, the *character associated with T*.

Thus a character is a function χ from G into the real numbers $\mathbf{R}$ or the complex numbers $\mathbf{C}$, depending upon whether we are thinking of T as being afforded by $\mathbf{R}^n$ or by $\mathbf{C}^n$. Let's consider first a simple example.

14.2   Proposition: The character of the regular representation

Let G be a finite group and let T be the right regular representation of G. Then the character χ associated with T is the function

$$\chi(1) = |G|, \qquad \chi(g) = 0 \text{ if } g \neq 1.$$

Proof: This follows immediately from the fact that the right regular representation is formed from a suitably arranged group table, as described in Chapter 11, and that $T(g)$ has a 1 precisely where g appears in that table; only the identity appears on the main diagonal. □

Clearly, if T is any one-dimensional representation of G, then T is related to its associated character χ by the equation

$$T(g) = [\chi(g)] \qquad \text{for } g \in G.$$

To study characters, we'll need to recall some ideas about the trace from linear algebra. In general, $tr(AB)$ is *not* equal to the product $tr(A)tr(B)$. However, the following result will give us all that we'll need in this direction.

14.3   Proposition: $tr(AB) = tr(BA)$

If A and B are $n \times n$ matrices, then $tr(AB) = tr(BA)$.

Proof: Let $A = (a_{ij})$ and $B = (b_{ij})$ with $1 \le i, j \le n$. Then,

$$tr(AB) = tr\left(\sum_{k=1}^{n} a_{ik}b_{kj}\right) = \sum_{i=1}^{n}\sum_{k=1}^{n} a_{ik}b_{ki}$$

$$= \sum_{k=1}^{n}\sum_{i=1}^{n} b_{ki}a_{ik} = tr\left(\sum_{i=1}^{n} b_{ki}a_{ij}\right) = tr(BA). \qquad □$$

Consequently, we have the following important results.

14.4   Proposition: $tr(A^{-1}BA) = tr(B)$

If A is nonsingular and if A and B both have degree n, then $tr(A^{-1}BA) = tr(B)$.

Proof: By Proposition 14.3,

$$tr(A^{-1}BA) = tr(A^{-1}(BA)) = tr((BA)A^{-1}) = tr(B(AA^{-1})) = tr(B). \quad □$$

14.5 Proposition: Equivalent representations have equal characters

If T and T^* are equivalent representations of a group G, then their associated characters are equal as functions.

Proof: Since T is equivalent to T^*, there exists a nonsingular matrix A such that $AT(g) = T^*(g)A$ for every $g \in G$. Then if $\chi(g) = tr(T(g))$ and $\chi^*(g) = tr(T^*(g))$ for each $g \in G$, we have

$$\chi(g) = tr(T(g)) = tr(A^{-1}T^*(g)A) = tr(T^*(g)) = \chi^*(g)$$

by Proposition 14.4, for every $g \in G$. □

Bear in mind that Proposition 14.5 says that χ and χ^* are equal *as functions*; that is, they have the same domain (namely, G) and they take the same value on each given element of that domain. Proposition 14.4 gives us another interesting property of group characters, as follows.

14.6 Proposition: A character is a class function

If χ is a character of a group G, then χ is a *class function*; that is, if x and y are conjugate in G, then $\chi(x) = \chi(y)$.

Proof: By hypothesis, there exists some $g \in G$ such that $g^{-1}xg = y$. Let T be the representation associated with the character χ. Then,

$$\chi(y) = tr(T(y)) = tr(T(g^{-1}xg)) = tr(T(g^{-1})T(x)T(g))$$
$$= tr(T(g)^{-1}T(x)T(g)) = tr(T(x)) = \chi(x),$$

using Proposition 14.4 at the next-to-last equivalence. □

Now from Definition 12.5, if a representation T is reducible, then T is equivalent to a representation T^* of the form

$$T^*(g) = \begin{bmatrix} U(g) & Q(g) \\ \mathbf{0} & V(g) \end{bmatrix}, \qquad g \in G,$$

where U and V are representations of G, $\mathbf{0}$ is a submatrix of zeros, and $Q(g)$ is a matrix (in general, not square) depending upon g. It is clear that the character of T^* is equal to the sum of the characters of U and V, and by Proposition 14.5, the character of T is also equal to the sum of the characters of U and V.

Conversely, just as we can form a new representation

$$T(g) = \begin{bmatrix} U(g) & \mathbf{0} \\ \mathbf{0} & V(g) \end{bmatrix}, \qquad g \in G,$$

from given representations U and V of G, so we can form a new character χ given characters λ and μ of G by setting

$$\chi(g) = \lambda(g) + \mu(g) \qquad \text{for all } g \in G.$$

Before we continue, a word of caution is in order. A representation is a homomorphism, and we have consistently used the fact that if T is a representation of G, then $T(xy) = T(x)T(y)$ for any $x, y \in G$. But (as was mentioned earlier), the trace function does not in general preserve products, and, thus, if χ is the character associated with T, it will not ordinarily be true that $\chi(xy)$ is equal to the product $\chi(x)\chi(y)$. Even when $y = x$, we will not necessarily have $\chi(x^2)$ equal to $\chi(x)^2$, as the example of

$$T(x) = \begin{bmatrix} 0 & 1 \\ 1 & 0 \end{bmatrix}$$

for the cyclic group $\langle x \rangle$ of order 2 shows.

Although we'll omit the proof, it is true that the converse of Proposition 14.5 holds when G is finite.

14.7 Theorem: Equal characters for finite G come from equivalent representations

Two representations of a *finite* group G are equivalent if their characters are equal as functions.

Now we'll return to recalling ideas from linear algebra. Let τ be a linear transformation from $\mathbf{C}^n$ to itself. If X is a nonzero vector in $\mathbf{C}^n$ and λ a (complex) scalar such that $\tau(X) = \lambda X$, then X is called a *characteristic vector* of τ and λ a *characteristic root* of τ. (Sometimes the terms *eigenvector* and *eigenvalue* are used.) If A is the matrix for τ (with respect to some basis for $\mathbf{C}^n$); that is, if $\tau(Y) = AY$ for each $Y \in \mathbf{C}^n$, then X and λ satisfying $AX = \lambda X$ are also called a characteristic vector and a character-istic root *of the matrix A.* If A is an $n \times n$ matrix, then one shows in linear algebra that A has n characteristic roots (not necessarily distinct), which are the n roots of the *characteristic polynomial* $\det(I_n x - A)$ of A, where x is regarded as the variable of the polynomial. When no confusion with the notation for the order of a group will result, we'll use the notation $|M|$ for $\det(M)$.

Examples that review these ideas are in the Exercises.

Now let $\alpha_1, \ldots, \alpha_n$ be the characteristic roots of the matrix A; then these are the roots of the characteristic polynomial of A. Hence,

$$|I_n x - A| = (x - \alpha_1)(x - \alpha_2) \cdots (x - \alpha_n)$$

$$= x^n - \left(\sum_{i=1}^{n} \alpha_i \right) x^{n-1} + \{\text{terms in } x^{n-2}, \text{etc.}\}.$$

Now we'll expand $|I_n x - A|$ by minors of the first row:

$$|I_n x - A| = \begin{vmatrix} x - a_{11} & -a_{12} & \cdots & -a_{1n} \\ -a_{21} & x - a_{22} & \cdots & -a_{2n} \\ & & \cdots & \\ -a_{n1} & -a_{n2} & \cdots & x - a_{nn} \end{vmatrix}$$

$$= (x - a_{11}) \begin{vmatrix} x - a_{22} & \cdots & -a_{2n} \\ & \cdots & \\ -a_{n2} & \cdots & x - a_{nn} \end{vmatrix} + Q,$$

where Q consists of terms involving only $(n-2)$ entries of the form $(x-a_{jj})$. Hence the coefficients of x^n and x^{n-1} in the characteristic polynomial must come only from the term preceding Q above. If we continue the process and also expand that term by minors of the first row (or use mathematical induction), we arrive at

$$|I_n x - A| = (x - a_{11})(x - a_{22}) \cdots (x - a_{nn}) + Q^*,$$

where Q^* consists of terms in x^{n-2}, etc. Comparing the coefficients of x^{n-1} with those in the first decomposition of $|I_n x - A|$, we see that

$$\sum_{j=1}^{n} \alpha_j = \sum_{j=1}^{n} a_{jj},$$

and we have now proved the following proposition.

14.8   Proposition: Trace and characteristic roots of A

The trace of a square matrix A equals the sum of the characteristic roots of A.

This result will be very helpful in calculating group characters. We'll also need to use facts about the powers and inverse of a matrix.

14.9   Proposition: Characteristic roots of A^k and A^{-1}

Let A be a matrix of degree n with characteristic roots $\alpha_1, \ldots, \alpha_n$, and let k be a positive integer. Then the characteristic roots of A^k are $\alpha_1^k, \ldots, \alpha_n^k$, and the characteristic roots of A^{-1} are $1/\alpha_1, \ldots, 1/\alpha_n$.

Proof: Let $1 \leq j \leq n$; then there exists a vector X_j such that $AX_j = \alpha_j X_j$. But then

$$A^k X_j = A^{k-1}(\alpha_j X_j) = \alpha_j A^{k-1} X_j$$
$$= \alpha_j A^{k-2}(\alpha_j X_j) = \alpha_j^2 A^{k-2} X_j,$$

and by repeating this process we arrive at

$$A^k X_j = \alpha_j^k X_j,$$

so α_j^k is a characteristic root of A^k (with associated characteristic vector X_j). Since A^k has only n characteristic roots, these roots must be the α_j^k for $1 \leq j \leq n$. For A^{-1}, from $AX_j = \alpha_j X_j$, we have

$$X_j = (A^{-1}A)X_j = A^{-1}(AX_j) = A^{-1}(\alpha_j X_j) = \alpha_j(A^{-1}X_j),$$

so $A^{-1}X_j = (1/\alpha_j)X_j$ and $1/\alpha_j$ is a characteristic root of A^{-1}. $\square$

14.10 Proposition: Characteristic roots of a matrix of finite order

If A is a matrix of finite order, then the characteristic roots of A are all roots of 1.

Proof: Let k be the order of A (meaning, as usual, the smallest positive integer k such that $A^k = I_n$, where n is the degree of A). Then by Proposition 14.9 the characteristic roots of A^k are the kth powers of the characteristic roots of A, but these must all equal 1 since all characteristic roots of I_n are equal to 1. □

For finite groups, Proposition 14.10 is of great importance because every element of a finite group has finite order and, since a representation is a homomorphism, every matrix used by the representation must have finite order.

Now recall that if $z = a + bi$ is a complex number with a, b real, then the *complex conjugate* (or simply *conjugate*) of z is $a - bi$, which is denoted by $\bar{z}$. We define the *norm* of z to be $\sqrt{z\bar{z}} = \sqrt{a^2 + b^2}$, which we denote as $|z|$. Note that if z is real, then $|z|$ is just the absolute value of z, and thus the notation $|z|$ is consistent. For any $z \in C$, $|z|$ is the length of the vector representing z in the complex plane, and $|z| = |\bar{z}|$. Important properties of the conjugate and of the norm follow, as review.

14.11 Proposition: Properties of the complex conjugate and norm

Let z_1 and $z_2 \in C$. Then,

$$\overline{z_1 + z_2} = \overline{z_1} + \overline{z_2},$$
$$\overline{z_1 z_2} = \overline{z_1}\,\overline{z_2},$$
$$\text{and} \quad |z_1 z_2| = |z_1|\,|z_2|.$$

The proofs are left to Exercise 14.4. Let's continue by considering kth roots of 1.

14.12   Proposition: The norm of a kth root of 1

If ζ is a kth root of 1, then $|\zeta| = 1$ and $\zeta^{-1} = \bar{\zeta}$.

Proof: By repeated applications of the third clause of Proposition 14.11, we have $|\zeta|^k = |\zeta^k| = |1| = 1$. But clearly, the norm of any complex number is a nonnegative real number, and the only nonnegative real root of 1 is 1 itself; hence $|\zeta| = 1$. Now since

$$\zeta\bar{\zeta} = |\zeta|\,|\zeta| = 1^2 = 1,$$

we have

$$\bar{\zeta} = 1/\zeta = \zeta^{-1}. \qquad \qquad \square$$

These concepts are applied to group characters by the following.

14.13   Proposition: Value of character on inverse of an element

If T is a representation of G with character χ, and if $x \in G$ is an element of finite order, then $\chi(x^{-1}) = \overline{\chi(x)}$.

Proof: Let k be the order of the matrix $T(x)$ in the group $GL(n, C)$. Then by Proposition 14.9, the characteristic roots of $T(x^{-1})$ are the inverses of those of $T(x)$, that is, the reciprocals, which are multiplicative inverses. But by Proposition 14.13, the inverse of a root of 1 is the complex conjugate. Now, by Proposition 14.8, $\chi(x)$ and $\chi(x^{-1})$ are, respectively, the sums of the characteristic roots of $T(x)$ and of $T(x^{-1})$. By $k - 1$ applications of Proposition 14.12, $\chi(x^{-1}) = \overline{\chi(x)}$. $\qquad \square$

Of course, if G is a finite group, then all of its elements have finite order, and Proposition 14.13 applies to all elements of G.

We'll use crucially, although without proof, the following result.

14.14   Theorem: Number of irreducible representations of finite G

The number of inequivalent irreducible representations of a finite group G over the complex field C is equal to the number of distinct conjugate classes of G.

As our first application of this important theorem, we'll prove the following.

14.15   Theorem: Irreducible representations of a finite abelian group

If G is a finite abelian group, then every irreducible representation of G has degree 1.

Proof: Every element of G is *self-conjugate*; that is, if $x \in G$, then $g^{-1}xg = x$ for every $x \in G$. Hence each element of G forms its own conjugate class, and Theorem 14.14 shows that the number of inequivalent irreducible representations of G is equal to $|G|$. But by Theorem 13.12, there are already $|G|$ one-dimensional representations of G, and these are certainly irreducible.   □

Let's return to the familiar example of D_4. Writing

$$D_4 = \langle x, y : x^4 = y^2 = 1, yx = x^{-1}y \rangle,$$

we can specify any k-dimensional representation of D_4 by giving $T(x)$ and $T(y)$ as $k \times k$ matrices and verifying that the defining relations

$$T(x)^4 = T(y)^2 = I_k \qquad \text{and} \qquad T(y)T(x) = T(x)^{-1}T(y)$$

hold. The one-dimensional representations of D_4 are found in Exercises 10.8, 13.11, and 14.5.

Now consider the representation of degree 2 given by

$$U(x) = \begin{bmatrix} 0 & -1 \\ 1 & 0 \end{bmatrix}, \qquad U(y) = \begin{bmatrix} 0 & 1 \\ 1 & 0 \end{bmatrix}.$$

These matrices do satisfy the defining relations of D_4 and so U is a representation. To show that U is irreducible over C, we need (by Maschke's theorem, Theorem 12.10) merely to show that U is indecomposable. Suppose, on the contrary, that U is decomposable. Then there exists

$$A = \begin{bmatrix} a & b \\ c & d \end{bmatrix}$$

with $ad - bc \neq 0$ and there exist complex numbers λ, μ, α, β such that

$$A^{-1}U(x)A = \begin{bmatrix} \lambda & 0 \\ 0 & \mu \end{bmatrix} \quad \text{and} \quad A^{-1}U(y)A = \begin{bmatrix} \alpha & 0 \\ 0 & \beta \end{bmatrix}.$$

Taking determinants on both sides, we have

$$1 = \lambda\mu \quad \text{and} \quad -1 = \alpha\beta.$$

Taking traces on both sides, we have

$$0 = \lambda + \mu \quad \text{and} \quad 0 = \alpha + \beta.$$

Hence $\lambda = \pm i$, $\mu = -\lambda$, $\alpha = \pm 1$, $\beta = -\alpha$. By Exercise 12.3, λ and μ are interchangeable, so we may simply set $\lambda = i = -\mu$. (A caution here: we do not assign one value or the other to α at this point because the interchange of λ and μ would force the same interchange of α and β.) Now rewrite the matrix products as

$$U(x)A = A \begin{bmatrix} i & 0 \\ 0 & -i \end{bmatrix} \quad \text{and} \quad U(y)A = A \begin{bmatrix} \alpha & 0 \\ 0 & \beta \end{bmatrix};$$

performing these products, we find that

$$\begin{bmatrix} -c & -d \\ a & b \end{bmatrix} = \begin{bmatrix} ia & -ib \\ ic & -id \end{bmatrix} \quad \text{and} \quad \begin{bmatrix} c & d \\ a & b \end{bmatrix} = \begin{bmatrix} \alpha a & \beta b \\ \alpha c & \beta d \end{bmatrix}.$$

Thus, $ia = -c$ and $\alpha a = c$, from which $\alpha a = -ia$, but $\alpha = \pm 1$, so $\alpha a = \pm a = -ia$, which implies that $a = 0$. But then, since $a = \alpha c$, it follows that $c = 0$ also. This makes $\det(A) = 0$, a contradiction; therefore, U is irreducible.

If χ is the character afforded by U, then (recalling Proposition 14.6) we can find $(U(x))^2$ and $U(x)U(y)$, then write $\chi(1) = 2$, $\chi(x) = 0$, $\chi(y) = 0$,

$\chi(x^2) = -2$, $\chi(xy) = 0$ as a list of the values taken by χ on elements of the distinct conjugate classes of D_4.

We'll combine all of this information into a *character table* for D_4. In such a table each column corresponds to a conjugate class (or to one representative of a conjugate class), and each row gives the values of one of the irreducible characters, which are the characters of the irreducible representations. Note that, by Theorem 14.14, a character table will be square.

It is customary to denote the irreducible characters of g by $\zeta^{(1)}$, $\zeta^{(2)}, \ldots,$ $\zeta^{(s)}$, where s is the number of distinct conjugate classes of G. We'll also always take $\zeta^{(1)}$ to be the trivial character of G, the one assigning the value 1 to every element of G.

Here, then, is the character table for D_4:

	Conjugate Classes				
Characters	$\{1\}$	$\{x^2\}$	$\{x, x^3\}$	$\{y, x^2y\}$	$\{xy, x^3y\}$
$\zeta^{(1)}$	1	1	1	1	1
$\zeta^{(2)}$	1	1	-1	1	-1
$\zeta^{(3)}$	1	1	1	-1	-1
$\zeta^{(4)}$	1	1	-1	-1	1
$\zeta^{(5)}$	2	-2	0	0	0

Exercises

14.1. Find the characteristic polynomial and the characteristic roots of the matrix
$$A = \begin{bmatrix} 1 & 3 \\ 2 & 2 \end{bmatrix}.$$

14.2. If A is a matrix of degree n and if $X \in \mathbf{C}^n$ is a characteristic vector of A, show that, for any nonzero complex scalar c, cX is also a characteristic vector for A.

14.3. Find characteristic vectors corresponding to the characteristic roots of the matrix A in Exercise 14.1.

14.4. Prove Proposition 14.11.

14.5. Find all of the one-dimensional representations of D_4.

14.6. Let g be an element of order 3 in a group G, let T be a representation of G with $\deg(T) = 2$ and associated character χ, and let ζ be a primitive cube root of 1. Find all six of the possible values that $\chi(g)$ might take on.

14.7. Let g be an element of order 4 in a group G, and let T be a representation of G with $\deg(T) = 3$ and associated character χ. Find all of the possible values that $\chi(g)$ might take.

14.8. Let g be an element of a group G with the property that g is conjugate to g^{-1}, and let χ be a character of G. Show that $\chi(g)$ must be a real number.

14.9. Let g be an element of order 2 in a group G, and let χ be a character of G. What can you conclude about $\chi(g)$?

14.10. If G is a group that coincides with its derived group G', what can you say about the one-dimensional characters of G?

14.11. The group $G = \langle a, b : a^5 = b^4 = 1, ba = a^2 b \rangle$ has order 20; its only nontrivial normal subgroup is $\langle a \rangle$. Find all of the one-dimensional representations of G by considering those of G/G'.

Chapter 15

Orthogonality Relations and Character Tables

We have seen that a character is a function from a group G into the complex numbers $\mathbf{C}$. If G is a *finite* group and if χ and ψ are *any* two functions from G into $\mathbf{C}$, we may define an inner product of these two characters as follows.

15.1 Definition: Inner product of complex-valued functions

Let χ and ψ be functions from a finite group G into the complex numbers $\mathbf{C}$. Define the *inner product* of χ and ψ by

$$\langle \chi, \psi \rangle = \frac{1}{|G|} \sum_{g \in G} \chi(g)\overline{\psi(g)},$$

where the bar represents complex conjugate (as in Chapter 14) and the summation indicates that there is to be one addend for each element of G (which makes sense since G is finite).

Notice in this definition that $\langle \chi, \psi \rangle$ is a complex number; this fact is very important for our work. This notation closely resembles that for a group with two generators, but the context will always make it clear which of the two interpretations is intended; in particular, we'll use lowercase Greek letters for characters and will continue to use lowercase Roman for elements.

As an example, let's consider the irreducible characters found for D_4 at the end of Chapter 14. Here

$$\langle \zeta^{(2)}, \zeta^{(5)} \rangle = \frac{1}{8}(2 - 2 + 0 + 0 + 0 + 0 + 0 + 0) = 0,$$

$$\langle \zeta^{(2)}, \zeta^{(2)} \rangle = \frac{1}{8}(1+1+1+1+1+1+1+1) = 1.$$

Before continuing with the discussion, let's justify the reference in Definition 15.1 to an inner product which, in the language of linear algebra, should be a complex-valued symmetric bilinear form.

15.2   Proposition: The inner product is a symmetric bilinear form

Let G be a finite group, let χ, ψ, and ϕ be functions from G into $\mathbf{C}$, and let $a, b \in \mathbf{C}$. Then, the inner product satisfies the conditions of *bilinearity*,

$$\langle a\chi + b\psi, \phi \rangle = a\langle \chi, \phi \rangle + b\langle \psi, \phi \rangle,$$

$$\langle \chi, a\psi + b\phi \rangle = \overline{a}\langle \chi, \psi \rangle + \overline{b}\langle \chi, \phi \rangle,$$

and of *symmetry*,

$$\langle \chi, \psi \rangle = \overline{\langle \psi, \chi \rangle}.$$

The proof of this result is routine and has been left as Exercises 15.1 and 15.2.

When the functions χ and ψ in Definition 15.1 are characters, we may use the fact (see Proposition 14.6) that characters are class functions to streamline the calculation. Since all elements of a given conjugate class have the same value for a given character, we may simply multiply that value by the number of elements in the conjugate class; doing so for each distinct conjugate class and summing the result will, of course, yield the same summation as in Definition 15.1. Thus, the following proposition.

15.3   Proposition: Inner product of group characters

Suppose that the finite group G has the distinct conjugate classes $C_1, \ldots,$ C_s, that class C_j contains h_j elements, and that $g_j \in C_j$ for $j = 1, \ldots, s$.

If χ and ψ are characters of G, then

$$\langle \chi, \psi \rangle = \frac{1}{|G|} \sum_{j=1}^{s} h_j \chi(g_j) . \overline{\psi(g_j)} .$$

Since D_4 has five conjugate classes, the calculation in Proposition 15.3 involves only five terms rather than eight, as in Definition 15.1. For larger groups, greater simplifications may come about, although for abelian groups the two expressions are identical (Why?).

Our next step will be to investigate the inner product of *irreducible characters* of a finite group, which are those associated with representations that are irreducible over the complex numbers $\mathbf{C}$.

Let δ_{ij} denote the *Kronecker delta* defined by

$$\delta_{ij} = \begin{cases} 1 & \text{if } i = j, \\ 0 & \text{if } i \neq j. \end{cases}$$

If you carry out the 25 computations of $\langle \zeta^{(i)}, \zeta^{(j)} \rangle$ for $1 \leq i, j \leq 5$ in D_4, you will find that $\langle \zeta^{(i)}, \zeta^{(j)} \rangle = \delta_{ij}$ for all i and j.

15.4 Theorem: Inner product of irreducible characters

Let G be a finite group having the distinct irreducible characters $\zeta^{(1)}, \ldots, \zeta^{(s)}$. Then,

$$\langle \zeta^{(i)}, \zeta^{(j)} \rangle = \delta_{ij}$$

for $1 \leq i, j \leq s$.

We won't prove this result, but we'll make extensive use of it.

The value of the ith irreducible character of a finite group G on the jth conjugate class of G is conventionally denoted by $\zeta_j^{(i)}$; specifically,

$$\zeta_j^{(i)} = \zeta^{(i)}(g_j), \qquad \text{where } g_j \in C_j.$$

Because of Theorem 14.14, both i and j will run over the integers $1, \ldots, s$, where s is the number of distinct conjugate classes of G.

Using this notation, we define a *character table* for a finite group G to be the matrix $(\zeta_j^{(i)})$, that is, the matrix whose ith row gives the values of the ith irreducible character for G (in some listing of those characters) and whose jth column gives the values those characters take on the jth conjugate class of G (again, in some listing). The table at the end of Chapter 14 is an example of a character table.

Now given a character table for a finite group G, we'll define the *conjugate* of any column (or row) of the table to be the column (or row) vector whose entries are, respectively, the complex conjugates of the entries in the given column (or row). Then from Theorem 15.4, we have the following corollary.

15.5   Corollary: Row products in a character table

From a character table for a finite group G, let X be the row vector $(h_j \zeta_j^{(i)})$ and $\bar{Y}$ be the conjugate of the row vector $(\zeta_j^{(k)})$. If $i \neq k$, then the ordinary dot product $X \cdot \bar{Y} = 0$. If $i = k$, then $X \cdot \bar{Y} = |G|$.

Proof:

$$X \cdot \bar{Y} = \sum_{j=1}^{s} (h_j \zeta_j^{(i)}) \overline{\zeta_j^{(k)}} = |G| \langle \zeta^{(i)}, \zeta^{(k)} \rangle = |G| \delta_{ij}.$$

In this corollary we have considered rows of a character table as vectors. Another theorem, which we'll also state without proof, will give a similar result for the columns.

15.6   Theorem: Products of irreducible characters

Let G be a finite group having the distinct irreducible characters $\zeta^{(1)}, \ldots, \zeta^{(s)}$, let $1 \leq i, j \leq s$, and let h_i denote the number of elements in the conjugate class C_i. Then

$$\sum_{k=1}^{s} h_i \zeta_i^{(k)} \overline{\zeta_j^{(k)}} = |G| \delta_{ij}.$$

From this theorem we can derive two useful corollaries about the columns of a character table. Notice that the ith and jth conjugate classes in Theorem 15.6 are fixed and that the sum is taken over the values of all of the irreducible characters on those two classes. Moreover, since i is fixed, the term h_i can be factored out of the sum, making the theorem equivalent to

$$\sum_{k=1}^{s} \zeta_i^{(k)} \overline{\zeta_j^{(k)}} = \frac{|G|}{h_i} \delta_{ij}.$$

Now $\zeta_i^{(k)}$ is the value of the character $\zeta^{(k)}$ on the elements of the ith conjugate class, which are the values found in the ith column of the character table, and similarly for $\zeta_j^{(k)}$. At this point we have actually proved the following.

15.7 Corollary: Product of columns in a character table

In a character table, the dot product of any column with the conjugate of any other column is 0, and the dot product of the ith column with its own conjugate is $|G|/h_i$.

At the end of Chapter 14 we remarked that we'll always take $\zeta^{(1)}$ to be the trivial character of G, the one assigning the value 1 to every element of G. Similarly, we'll always take $C_1 = \{1\}$; then the first column of a character table will give the trace of the identity matrices whose degrees are the degrees of the irreducible representations associated with the characters. In other words, the first column of a character table will give the degree of the irreducible characters of the group. Since $h_1 = 1$, we have now proved the following.

15.8 Corollary: Degrees of the irreducible characters

The sum of the squares of the distinct irreducible characters of a finite group G is equal to $|G|$.

Corollaries 15.5, 15.7, and 15.8 are very powerful tools for working out the character table for a finite group. Let's look at some examples.

15.9   Example: The character table for S_3

The nonabelian group S_3 (equivalently, D_3) of order 6 has three conjugate classes, specifically,

$$C_1 = \{1\},$$
$$C_2 = \{(12), (13), (23)\},$$
$$C_3 = \{(123), (132)\}.$$

Now Corollary 15.8 tells us that we need integers z_1, z_2, z_3 such that $z_1^2 + z_2^2 + z_3^2 = 6$. You can easily see that the only solution is two 1 integers and a 2. It is customary to list the irreducible characters in a table in order of increasing degree, so we'll take $z_1 = z_2 = 1$ and $z_3 = 2$. Now, of course, we have the trivial representation:

$$T_1(g) = [1] \qquad \text{for all } g \in S_3,$$

and it gives the *trivial character*

$$\zeta^{(1)}(g) = 1 \qquad \text{for all } g \in S_3.$$

In Chapter 10 we found a nontrivial one-dimensional representation of S_3, which gives the character $\zeta_1^{(2)} = 1$, $\zeta_2^{(2)} = -1$, $\zeta_3^{(2)} = 1$. The information we have so far may be put into the character table as

	C_1	C_2	C_3
$\zeta^{(1)}$	1	1	1
$\zeta^{(2)}$	1	−1	1
$\zeta^{(3)}$	2		

Now compare the second column with the first and then the third column with the first using Corollary 15.7; this gives the values for $\zeta_2^{(3)}$ and $\zeta_3^{(3)}$ and allows us to complete the table as

	C_1	C_2	C_3
$\zeta^{(1)}$	1	1	1
$\zeta^{(2)}$	1	−1	1
$\zeta^{(3)}$	2	0	−1

In the last step of the example, we took the product of the second column with the first; the reason for taking the terms in this order is that the first column is known to consist entirely of real numbers (in fact, integers), and this allowed us to dispense with the complex conjugates (which are troublesome to remember).

15.10   Example: The character table for Q_2

Next we'll look at the character table for the quaternion group, of order 8. An easy computation shows that the conjugate classes of Q_2 are

$$C_1 = \{1\}, \qquad C_2 = \{-1\}, \qquad C_3 = \{\pm i\}, \qquad C_4 = \{\pm j\}, \qquad C_5 = \{\pm k\}.$$

You found the four one-dimensional representations of Q_2 in Exercise 10.9. Now from Corollary 15.8, we have

$$1^1 + 1^2 + 1^2 + 1^2 + z_5^2 = 8,$$

so $z_5 = 2$. This much information provides most of the character table:

	C_1	C_2	C_3	C_4	C_5
$\zeta^{(1)}$	1	1	1	1	1
$\zeta^{(2)}$	1	1	−1	1	−1
$\zeta^{(3)}$	1	1	1	−1	−1
$\zeta^{(4)}$	1	1	−1	−1	1
$\zeta^{(5)}$	2				

To complete the values of $\zeta^{(5)}$, apply Corollary 15.7 to the second through fifth columns, each in a product with the first. The final row now reads

$$\zeta^{(5)} \mid \quad 2 \quad -2 \quad 0 \quad 0 \quad 0$$

Compare this table with the one found for D_4 at the end of Chapter 14: the two are identical even though the groups are not isomorphic. Thus we cannot hope that a group will be characterized by its character table alone.

Now let's work out the character table for the alternating group A_4 of order 12.

15.11   Example: The character table for A_4

The conjugate classes of A_4 are

$$C_1 = \{1\},$$
$$C_2 = \{(12)(34), (13)(24), (14)(23)\},$$
$$C_3 = \{(123), (142), (134), (243)\},$$
$$C_4 = \{(132), (124), (143), (234)\}.$$

Now you can easily check that $H = C_1 \cup C_2$ is a subgroup of A_4. Moreover, H consists of all of the elements of orders 1 and 2 in A_4, and since conjugation preserves the orders of elements, this shows that H is a normal subgroup of A_4. Since $A_4/H \cong Z_3$, we know from Chapter 13 that H is the derived group and that A_4 has precisely three one-dimensional representations. But since there are exactly four conjugate classes, there are exactly four irreducible characters, and

$$1^2 + 1^2 + 1^2 + z_4^2 = 12,$$

so the fourth irreducible character has degree 3. From orthogonality of the columns, the character table can be completed as follows:

	C_1	C_2	C_3	C_4
$\zeta^{(1)}$	1	1	1	1
$\zeta^{(2)}$	1	1	ζ	ζ^2
$\zeta^{(3)}$	1	1	ζ^2	ζ
$\zeta^{(4)}$	3	-1	0	0

where ζ is a primitive cube root of 1. In checking the orthogonality of the columns, you need to use the identity

$$1 + \zeta + \zeta^2 = \frac{1 - \zeta^3}{1 - \zeta} = \frac{0}{1 - \zeta} = 0.$$

Since H is the derived group of A_4, the characters of degree 1 have all entries of 1 in the first two columns, reflecting Proposition 13.16. The values in the last two columns come from the isomorphism of A_4/H with Z_3, using Proposition 13.13.

These examples were relatively easy to carry out, so we need to look at something a bit more difficult.

15.12   Example: The character table for S_4

The symmetric group S_4 has order 24, and its derived group is A_4. Since $[S_4 : A_4] = 2$, Theorem 13.17 says that S_4 has exactly two characters of degree 1, of which $\zeta^{(1)}$ is (as usual) the trivial character. By computation you can determine the conjugate classes of S_4 (see Exercises 15.4 and 15.5); they may be tabulated as follows:

j	h_j	Description of C_j
1	1	$\{1\}$
2	6	elements of the form $(\alpha\beta)$
3	8	elements of the form $(\alpha\beta\gamma)$
4	6	elements of the form $(\alpha\beta\gamma\delta)$
5	3	elements of the form $(\alpha\beta)(\gamma\delta)$

Here α, β, γ, δ take on the values 1, 2, 3, 4 in some order and are distinct; as in the past, $h_j = |C_j|$. Observe that the derived group A_4 consists of $C_1 \cup C_3 \cup C_5$; by Proposition 13.13, $\zeta_2^{(2)} = \zeta_4^{(2)} = -1$ from the nontrivial character of Z_2. Corollary 15.8 says that

$$1^2 + 1^2 + z_3^2 + z_4^2 + z_5^2 = 24;$$

a bit of experimentation shows that $z_3 = 2$ and $z_4 = z_5 = 3$. The subgroup $H = C_1 \cup C_5$ turns out to be normal in S_4, as it was in A_4 (see Exercise 15.7), but it does not contain the derived group A_4 and so must be the nonabelian group of order 6. But in Example 15.9 we found the character table for this group, so Proposition 13.13 allows us to work out the irreducible character of degree 2 for S_4, as follows. In the factor group (which is isomorphic to S_3), $H(12)$ and $H(1234)$ have order 2 and hence correspond to the elements in the second conjugate class of S_3; thus (from the earlier character table) $\zeta_2^{(3)} = \zeta_4^{(3)} = 0$. Similarly, $H(123)$ has order 3 and corresponds to the third conjugate class of S_3; thus $\zeta_3^{(3)} = -1$. The value $\zeta_5^{(3)} = 2$ comes from the fact that $C_5 \subseteq H$. Incorporating what we have found thus far into the table, we have

	C_1	C_2	C_3	C_4	C_5
$\zeta^{(1)}$	1	1	1	1	1
$\zeta^{(2)}$	1	−1	1	−1	1
$\zeta^{(3)}$	2	0	−1	0	2
$\zeta^{(4)}$	3				
$\zeta^{(5)}$	3				

Now apply Corollary 15.7 to the second through fourth columns with the first; this gives

$$\zeta_2^{(5)} = -\zeta_2^{(4)}, \qquad \zeta_3^{(5)} = -\zeta_3^{(4)}, \qquad \zeta_4^{(5)} = -\zeta_4^{(4)}.$$

For the fifth column with the first, we have

$$(1 \cdot 1) + (1 \cdot 1) + (2 \cdot 2) + 3\zeta_5^{(4)} + 3\zeta_5^{(5)} = 0,$$

whence $\zeta_5^{(5)} = -2 - \zeta_5^{(4)}$. Assigning variables to the unknown entries in the fourth row and using the equations just found, we can write the last two rows as

$$
\begin{array}{c|ccccc}
\zeta^{(4)} & 3 & t & u & v & w \\
\zeta^{(5)} & 3 & -t & -u & -v & -2 - w
\end{array}
$$

Apply the second part of Corollary 15.7 to the third column:

$$(\text{col } 3) \cdot \overline{(\text{col } 3)} = 3 + 2u\bar{u} = 24/8 = 3$$

so $u\bar{u} = 0$ and thus $u = 0$. Working with the second column, the elements in this conjugate class have order 2 and so are self-inverse; hence, from Proposition 14.13 t must be real (it must be equal to its own complex conjugate). This gives

$$(\text{col } 2) \cdot \overline{(\text{col } 2)} = 2 + 2t^2 = 24/6 = 4,$$

so $t^2 = 1$ and $t = \pm 1$. Since $\zeta_2^{(5)} = -\zeta_2^{(4)}$, we may simply set $t = 1$. Then, from $(\text{col } 4) \cdot \overline{(\text{col } 2)} = 2 + 2v = 0$ we obtain $v = -1$, and from $(\text{col } 5) \cdot \overline{(\text{col } 2)} = 0$ comes $w = -1$. This completes the calculation of the character table for S_4; in writing down the final result, we'll include a row with the values for $h_j = |C_j|$ since these numbers will be useful for later reference.

	C_1	C_2	C_3	C_4	C_5
h_j	1	6	8	6	3
$\zeta^{(1)}$	1	1	1	1	1
$\zeta^{(2)}$	1	-1	1	-1	1
$\zeta^{(3)}$	2	0	-1	0	2
$\zeta^{(4)}$	3	1	0	-1	-1
$\zeta^{(5)}$	3	-1	0	1	-1

Anyone who is skeptical of our choice of $t = 1$ can now verify that choosing $t = -1$ would merely interchange $\zeta^{(4)}$ with $\zeta^{(5)}$.

Notice the power of the orthogonality relations: we were able to calculate the two irreducible characters of degree 3 without ever looking at a matrix!

Having worked out the character tables for S_3, A_4, and S_4, we would naturally try next for A_5. Unfortunately, this project turns out to be significantly more difficult, and we won't be able to complete it until we reach the concept of induced characters in Chapter 19. At this point, however, we might as well put down as much information as we readily can.

The alternating group A_5 of order 60 (which we studied at length in Chapter 8) is simple and nonabelian; hence, the only character of degree 1 is the trivial character $\zeta^{(1)}$. The conjugate classes for A_5 may be tabulated as follows:

j	h_j	Description of C_j
1	1	$\{1\}$
2	15	elements of the form $(\alpha\beta)(\gamma\delta)$
3	20	elements of the form $(\alpha\beta\gamma)$
4	12	conjugates of (12345)
5	12	conjugates of (12354)

Borrowing just one piece of information from the next section, we'll find there an irreducible character of degree 4. Hence, the sum of the squares of the degrees of the other three irreducible characters must be $60 - (1^2 + 4^2) = 43$. Clearly, the maximum degree is 6 (since $7^2 = 49$ is already too large), and in fact it is easy to check that $a^2 + b^2 + c^2 = 43$ has the unique solution in integers 3, 3, 5. The information so far is pretty meager; here it is

	C_1	C_2	C_3	C_4	C_5
h_j	1	15	20	12	12
$\zeta^{(1)}$	1	1	1	1	1
$\zeta^{(2)}$	3				
$\zeta^{(3)}$	3				
$\zeta^{(4)}$	4				
$\zeta^{(5)}$	5				

Surely an attack by brute force, introducing 16 variables and applying the orthogonality relations, is not a desirable approach.

Exercises

15.1. Verify the first of the two equations for bilinearity in Proposition

15.2.

15.2.   Show that if $z_1, z_2 \in \mathbf{C}$, then $z_1 \overline{z_2} = \overline{z_2 \overline{z_1}}$. Use this fact as a lemma to verify the condition of symmetry in Proposition 15.2. Then use symmetry to check the second equation for bilinearity.

15.3.   Note that two distinct conjugate classes of A_4 unite into a single conjugate class of S_4. Why is this possible?

15.4.   Verify that all of the elements of the form $(\alpha\beta)$ in S_4 are indeed conjugate. To do so, first explain why it suffices to consider the two pairs $(\alpha\beta)$, $(\alpha\gamma)$ and $(\alpha\beta)$, $(\gamma\delta)$, showing that the elements of each pair are conjugate. (Here α, β, γ, δ represent distinct choices of points 1, 2, 3, 4.) Then find elements x and y of S_4 (in terms of the variables α, β, γ, δ) for which

$$x^{-1}(\alpha\beta)x = (\alpha\gamma) \qquad \text{and} \qquad y^{-1}(\alpha\beta)y = (\gamma\delta).$$

Finally, show that no element of order 2 that is not a 2-cycle can be conjugate to any $(\alpha\beta)$.

15.5.   Use simple combinatoric arguments (as was done for A_5 in Chapter 8) to check the values given for $h_1, \ldots, h_5$ in Example 15.12.

15.6.   A subgroup of a group G is called a *characteristic subgroup* if, whenever σ is an automorphism of G, then $\sigma(H) = H$. Prove that a characteristic subgroup is always normal, but find a specific example to show that a normal subgroup need not be characteristic.

15.7.   Let $H \leq G$, and assume that H contains all of the elements of G having one or more specified orders, and no other elements. You have assumed that H is a subgroup; prove that it is characteristic. (This shows that the subgroup $H = C_1 \cup C_5$ is normal in S_4.)

15.8.   Show that the assumption that H is a subgroup is essential in Exercise 15.7 by finding a specific example of a group having a subset consisting of the identity element and all elements of a given order that is not a subgroup.

15.9.   Prove that the center $C(G)$ of a group G and the derived group G' are characteristic subgroups of G.

15.10.   Find the conjugate classes and character table of the dihedral group of order 10 (the symmetries of the regular pentagon). *Suggestion:* Use Propositions 14.8 and 14.10.

Chapter 16

Reducible Characters

By Maschke's theorem (12.10), every complex representation of a finite group G is completely reducible. If T is any complex representation of G and if χ is the character associated with T, then T is equivalent to a representation of the form:

$$\begin{bmatrix} T_1(g) & & & \\ & T_2(g) & & \mathbf{0} \\ & & \ddots & \\ \mathbf{0} & & & T_r(g) \end{bmatrix}$$

where $T_1, \ldots, T_r$ are irreducible and $g \in G$. Hence by Proposition 14.4, χ is equal to a linear combination of the irreducible characters $\zeta^{(1)}, \ldots, \zeta^{(s)}$ of G with nonnegative integers as coefficients; that is, given χ, there exist integers $a_1, \ldots, a_s \geq 0$, such that

$$\chi = \sum_{i=1}^{s} a_i \zeta^{(i)}.$$

This expression means a sum of functions; specifically, it means that

$$\chi(g) = \sum_{i=1}^{s} a_i \zeta^{(i)}(g)$$

for every $g \in G$. For example, you can check that the regular representation of V_4 is the sum of the four irreducible characters of V_4.

Let's expand this idea a bit. Suppose that χ is a character of G with

$$\chi = \sum_{i=1}^{s} a_i \zeta^{(i)}.$$

Then, for a given irreducible character $\zeta^{(j)}$ of G,

$$\langle \chi, \zeta^{(j)} \rangle = \frac{1}{|G|} \sum_{k=1}^{s} h_k \chi(g_k) \overline{\zeta_k^{(j)}}$$

in the notation of Chapter 15, where C_k is a conjugate class of G, $g_j \in C_j$, and $h_k = |C_k|$. But we can substitute the expression for χ as a linear combination of irreducible characters directly into $\langle \chi, \zeta^{(j)} \rangle$ and obtain

$$\langle \chi, \zeta^{(j)} \rangle = \left\langle \sum_{i=1}^{s} a_i \zeta^{(i)}, \zeta^{(j)} \right\rangle$$

$$= \sum_{i=1}^{s} a_i \langle \zeta^{(i)}, \zeta^{(j)} \rangle \quad \text{by Proposition 15.2}$$

$$= \sum_{i=1}^{s} a_i \delta_{ij} \quad \text{by Theorem 15.4}$$

$$= a_j.$$

Thus, we have shown that $\langle \chi, \zeta^{(j)} \rangle = a_j$ for each $j = 1, \ldots, s$, and we have proved the following.

16.1   Theorem: Reduction of a character of G

Let G be a finite group having irreducible characters $\zeta^{(1)}, \ldots, \zeta^{(s)}$, and let χ be an arbitrary character of G. Then,

$$\chi = \sum_{i=1}^{s} \langle \chi, \zeta^{(i)} \rangle \zeta^{(i)}.$$

This result is very powerful; it can be used to reduce an arbitrary character χ of a finite group to the sum of the irreducible characters of which χ is composed, that is, to the sum of its irreducible components, and it is accomplished by a simple calculation.

16.2 Definition: Multiplicity and constituent

In the notation of Theorem 16.1, we call $a_i = \langle \chi, \zeta^{(i)} \rangle$ the *multiplicity* of $\zeta^{(i)}$ in χ; when $a_i > 0$, $\zeta^{(i)}$ is called a *constituent* of χ.

It turns out that the regular representation of a finite group has a striking characterization as the sum of irreducible characters.

16.3 Proposition: The character of the regular representation

If G is a finite group and χ is the character of the regular representation of G, then the multiplicity of each irreducible character of G in χ is equal to the degree of that character.

Proof: In Proposition 14.2 you saw that $\chi(1) = |G|$ and that $\chi(g) = 0$ for all $g \neq 1$. As usual, we'll take $C_1 = \{1\}$; then $\chi(g_j) = 0$ for all $g_j \in C_j$ with $j > 1$. Hence,

$$\langle \chi, \zeta^{(i)} \rangle = \frac{1}{|G|} \sum_{j=1}^{s} h_j \chi(g_j) \overline{\zeta_j^{(i)}}$$

$$= \frac{1}{|G|} \chi(1) \overline{\zeta_1^{(i)}}$$

$$= \frac{1}{|G|} \cdot |G| \cdot z_i$$

$$= z_i,$$

where (as usual) $z_i = \zeta_1^{(i)}$ is the degree of $\zeta^{(i)}$, as was to be proved. □

This result echoes Corollary 15.8: the degree of $\zeta^{(i)}$ and its multiplicity are equal, their product is z_i^2, and

$$\sum_{i=1}^{s} z_i^2 = |G|,$$

which is the degree of the regular representation.

16.4   Example: Reduction of a character of D_4

In Exercise 10.4, you found a representation of D_4 given by

$$U(r) = \begin{bmatrix} 0 & -1 & 0 & 0 \\ 1 & 0 & 0 & 0 \\ 0 & 0 & 0 & -1 \\ 0 & 0 & 1 & 0 \end{bmatrix}, \qquad U(c) = \begin{bmatrix} 0 & 1 & 1 & 0 \\ 1 & 0 & 0 & -1 \\ 0 & 0 & 0 & 1 \\ 0 & 0 & 1 & 0 \end{bmatrix}.$$

If χ is the character of U, then $\chi(r) = 0$ and $\chi(c) = 0$. Now remember that, when $\deg(\chi)$ is greater than 1, we cannot find $\chi(r^2)$ and $\chi(rc)$ as the products $\chi(r)^2$ and $\chi(r)\chi(c)$. But by matrix multiplication we obtain

$$U(r^2) = \begin{bmatrix} -1 & 0 & 0 & 0 \\ 0 & -1 & 0 & 0 \\ 0 & 0 & -1 & 0 \\ 0 & 0 & 0 & -1 \end{bmatrix}, \qquad U(rc) = \begin{bmatrix} -1 & 0 & 0 & 1 \\ 0 & 1 & 1 & 0 \\ 0 & 0 & -1 & 0 \\ 0 & 0 & 0 & 1 \end{bmatrix},$$

from which we have $\chi(r^2) = -4$ and $\chi(rc) = 0$. Comparison with the character table found at the end of Chapter 14 makes it immediately clear that $\chi = 2\zeta^{(5)}$, but it may be helpful for you to verify directly that $\langle \chi, \zeta^{(i)} \rangle = 0$ for $i = 1, 2, 3, 4$ (see Exercise 16.1).

Now let's see how this technique might apply to a more complicated example (even though it is purely hypothetical). Suppose that S_4 has the following character:

χ	C_1	C_2	C_3	C_4	C_5
	19	5	-2	-3	3

Then, using the character table we found at the end of Chapter 15, we find that

$$\langle \chi, \zeta^{(1)} \rangle = \frac{1}{24}(19 + 30 - 16 - 18 + 9) = 1,$$

$$\langle \chi, \zeta^{(2)} \rangle = \frac{1}{24}(19 - 30 - 16 + 18 + 9) = 0,$$

$$\langle \chi, \zeta^{(3)} \rangle = \frac{1}{24}(38 + 0 + 16 + 0 + 18) = 3,$$

$$\langle \chi, \zeta^{(4)} \rangle = \frac{1}{24}(57 + 30 + 0 + 18 - 9) = 4,$$

$$\langle \chi, \zeta^{(5)} \rangle = \frac{1}{24}(57 - 30 + 0 - 18 - 9) = 0,$$

from which we have the decomposition $\chi = \zeta^{(1)} + 3\zeta^{(3)} + 4\zeta^{(4)}$, which you can easily verify by writing out the values of this linear combination.

A more interesting example for S_4 may be derived from the group of rigid symmetries of the cube once we prove that the two groups are isomorphic.

16.5   Proposition: The group of the cube is isomorphic to S_4

The group of rigid symmetries of the cube is isomorphic to S_4.

Proof: Let the vertices of a cube be numbered as shown in Figure 10.1 (Chapter 10), and denote the four principal diagonals (the interior diagonals, each having length $\sqrt{3}$ times that of an edge) as

$$a = \overline{17}, \qquad b = \overline{28}, \qquad c = \overline{35}, \qquad d = \overline{46}.$$

Because we're considering only rigid symmetries of the cube, any such motion permutes the four diagonals among themselves, and since $|G| = 24$, all of the permutations of the set $\{a, b, c, d\}$ of the diagonals must be included. This assertion depends upon the observation that no two elements of the symmetry group G of the cube induce the same permutation of the diagonals, which may be verified as indicated in Exercise 16.2. $\qquad \square$

For future reference, we'll make this isomorphism explicit by giving corresponding elements of the conjugate classes.

16.6   Example: A correspondence of the cubic group to S_4

The elements of the group of rigid symmetries of the cube correspond under isomorphism to the elements of the permutation group S_4 as follows:

Elements of the Cubic Group	Elements of S_4	h_j
e	1	1
$r = (1234)(5678)$	$(abcd) \in C_4$	6
$h = (245)(386)$	$(bdc) \in C_3$	8
$r^2 = (13)(24)(57)(68)$	$(ac)(bd) \in C_5$	3
$rh = (14)(28)(35)(67)$	$(ad) \in C_2$	6

Here we've taken the conjugate classes of S_4 as we gave them in Example 15.12.

16.7   Example: A permutation representation for S_4

Just as we produced the (right) regular representation of a finite group by identifying the elements of the group with the natural basis vectors of a vector space, so we can produce a permutation representation of S_4 by identifying the natural basis vectors $e_1, \ldots, e_8$ with the vertices of the cube, taking advantage of the isomorphism we have just found. To be specific, let e_i correspond to vertex i of the cube, as the solid is shown in Figure 10.1; for $g \in S_4$, let $T(g)$ take vector e_i to vector e_j precisely when g takes vertex i to vertex j, and let χ be the character associated with T. Now $T(g)$ has a 1 in the (i,j)-entry precisely when g takes i to j, and 0 in all other positions. The character $\chi(g)$, being the trace of the matrix $T(g)$, simply counts the number of times 1 appears on the main diagonal, which is to say that $\chi(g)$ counts the number of vertices left fixed by g. It should be clear that T is indeed a representation of S_4. Using the elements given in Example 16.6, we immediately have the values of χ:

	C_1	C_2	C_3	C_4	C_5
χ	8	0	2	0	0

Let's use Theorem 16.1 to reduce χ to a sum of the irreducible characters of S_4, which were found in Example 15.12:

$$\langle \chi, \zeta^{(1)} \rangle = \frac{1}{24}(8 + 0 + 16 + 0 + 0) = 1,$$

$$\langle \chi, \zeta^{(2)} \rangle = \frac{1}{24}(8 + 0 + 16 + 0 + 0) = 1,$$

$$\langle \chi, \zeta^{(3)} \rangle = \frac{1}{24}(16 + 0 - 16 + 0 + 0) = 0,$$

$$\langle \chi, \zeta^{(4)} \rangle = \frac{1}{24}(24 + 0 + 0 + 0 + 0) = 1,$$

$$\langle \chi, \zeta^{(5)} \rangle = \frac{1}{24}(24 + 0 + 0 + 0 + 0) = 1.$$

We have shown that $\chi = \zeta^{(1)} + \zeta^{(2)} + \zeta^{(4)} + \zeta^{(5)}$. If you had the time and the patience, you could write down matrices for T and reduce the representation to its irreducible components, but this example does the job much more easily and shows the power of using characters.

Back in Chapter 10 we found a representation of S_4 that was given by

$$T(r) = \begin{bmatrix} 0 & -1 & 0 \\ 1 & 0 & 0 \\ 0 & 0 & 1 \end{bmatrix} \quad \text{and} \quad T(h) = \begin{bmatrix} 0 & -1 & 0 \\ 0 & 0 & -1 \\ 1 & 0 & 0 \end{bmatrix}.$$

(The element of S_4 denoted by h in this section was called d in Chapter 10; the notation is changed here because d was used for one of the diagonals in Proposition 16.5 and Example 16.6.) Let's denote the character of T by ψ. Then from the correspondence in Example 16.6, the values of ψ may be tabulated as

	C_1	C_2	C_3	C_4	C_5
ψ	3	-1	0	1	-1

You can see immediately that $\psi = \zeta^{(5)}$, even without doing the reduction by the method of Theorem 16.1.

At the end of Chapter 15 we got a small start on finding the character table for A_5; we'll conclude this section by finding a nontrivial irreducible character for this group.

16.8   Example: An irreducible character for A_5

The procedure here will be exactly the same as that in Example 16.7, using the conjugate classes of A_5 found at the end of Chapter 15. For each $g \in A_5$, let $T(g)$ be the permutation matrix having 1 as its (i, j)-entry when $g : i \rightarrow j$ (where i and j are points in the set $\{1, 2, 3, 4, 5\}$) and 0 in all other entries. Then T is a representation of A_5, and its associated character χ simply counts the number of points left fixed by any element in a given conjugate class. For example, $\chi((12)(34)) = 1$, and so $\chi(g_2) = 1$

for $g_2 \in C_2$. Thus χ is given as

	C_1	C_2	C_3	C_4	C_5
χ	5	1	2	0	0

Of course, we cannot assume that χ is irreducible (and, in fact, it is not). Thus far the only irreducible character we have for A_5 is the trivial character $\zeta^{(1)}$. Now

$$\langle \chi, \zeta^{(1)} \rangle = \frac{1}{60}(1 \cdot 5 + 15 \cdot 1 + 20 \cdot 2 + 0 + 0) = 1;$$

thus the multiplicity of $\zeta^{(1)}$ in χ is 1. Write χ as $\chi = \zeta^{(1)} + \psi$, where (so far as we know now) ψ might or might not be irreducible. But if ψ is reducible, then it must be the sum of two characters (at least one of which may be assumed to be irreducible) of smaller degree, and since ψ has degree 4, the sum of the degrees of those characters must be either $1 + 3$ or $2 + 2$. But $\zeta^{(1)}$ is not a constituent of ψ (if it were, then its multiplicity in χ would be greater than 1), so the question is whether or not A_5 has an irreducible character of degree 2 (remember that the table at the end of Chapter 15 borrowed the fact that A_5 has an irreducible character of degree 4 from this section!). Suppose that A_5 does have an irreducible character of degree 2; then by Corollary 15.8, the remaining irreducible characters have degrees satisfying the equation:

$$1^2 + 2^2 + z_3^2 + z_4^2 + z_5^2 = 60.$$

This means that we need a solution in integers to the equation $z_3^2 + z_4^2 + z_5^2 = 55$. We can (as usual) take $z_3 \leq z_4 \leq z_5$. If z_5 is 5, 6, or 7, we obtain $z_3^2 + z_4^2 = 30$, 19, or 6, none of which is the sum of two squares. On the other hand, if $z_5 \leq 4$, then $z_3^2 + z_4^2 + z_5^2 \leq 48$. The conclusion is that ψ is an irreducible character of degree 4, tabulated as

	C_1	C_2	C_3	C_4	C_5
ψ	4	0	1	-1	-1

We'll be able to complete the character table for A_5 once we have the construction of a character of a group from that of a subgroup (induced characters, Chapter 19).

Exercises

16.1. Verify that $\langle \chi, \zeta^{(i)} \rangle = 0$ when $1 \leq i \leq 4$ for the character χ of D_4 considered in Example 16.4.

16.2. Continue the exploration of the isomorphism in Proposition 16.5 as follows. The orbit of a vertex, say 1, contains all eight points of the set of vertices. What is the orbit of the diagonal $\overline{17}$ under the group of the cube? What is the stabilizer of a diagonal of the cube? Now apply Theorem 3.13 to the permutation group on the diagonals of the cube.

16.3. Follow the procedure of Example 16.7, using the vectors e_1, e_2, e_3, e_4 in $\mathbf{C}^4$ as corresponding to the diagonals a, b, c, d of the cube as in Proposition 16.5 and Example 16.6. Define a permutation representation on these diagonals, find its character, and reduce this character to the sum of its irreducible components.

16.4. Repeat Exercise 16.3, using instead the six faces of the cube (see Figure 10.1):

$$A = \overline{1234} \qquad C = \overline{1485} \qquad E = \overline{3487}$$
$$B = \overline{1265} \qquad D = \overline{2376} \qquad F = \overline{5678}.$$

Define a permutation representation on these faces, find its character, and reduce this character to the sum of its irreducible components.

16.5. Repeat Exercise 16.3, using instead the 12 edges of the cube. Define a permutation representation on these edges, find its character, and reduce this character to the sum of its irreducible components.

16.6. Repeat Exercise 16.3, using instead the vertices, faces, and edges of the regular tetrahedron (Figure 3.2). Define permutation representations on these sets, find their characters, and reduce each character to the sum of its irreducible components.

16.7. Consider the group

$$H = \langle r, s : r^5 = s^4 = 1, rs = sr^2 \rangle.$$

First, show that, for any nonnegative integers i, j, k,

$$s^i r^j s^k = s^{i+1} r^{2j} s^{k-1}.$$

Then use this fact to show that every element of H can be written in the form $s^i r^j$. Finally, show that H has order 20.

16.8.   Show that $\sigma(a) = r$, $\sigma(b) = s^3$ determines an isomorphism of the group G of Exercise 14.11 onto the group H of Exercise 16.7.

16.9.   Find the five conjugate classes and the character table of the group (G or H) of Exercise 16.8.

16.10.   Show that

$$U(r) = I_3, \qquad U(s) = \begin{bmatrix} 1 & 0 & -1-i \\ -1+i & -i & 1-i \\ 0 & 0 & -i \end{bmatrix}$$

determines a representation of the group of Exercise 16.9. Find its character, and reduce that character to its irreducible components.

16.11.   One of the nonabelian groups of order 16 (there are nine in all) is isomorphic to $D_4 \times Z_2$. Find its conjugate classes and its character table.

16.12.   Another nonabelian group of order 16 is

$$G = \langle x, y : x^8 = y^2 = 1, yx = x^5 y \rangle.$$

Show that, if ζ is a primitive eighth root of 1, then

$$T(x) = \begin{bmatrix} \zeta & 0 \\ 0 & \zeta^5 \end{bmatrix}, \qquad T(y) = \begin{bmatrix} 0 & 1 \\ 1 & 0 \end{bmatrix}$$

determines a representation of G. Show that $C(G) = \langle x^2 \rangle$. Then find the conjugate classes and the character table of G.

16.13.   Find the character table of D_6, the dihedral group of order 12.

16.14.   There are three nonabelian groups of order 18. Two are the dihedral group D_6 and the direct product $S_3 \times Z_3$; the third is

$$G = \langle a, b, c : a^3 = b^3 = c^2 = 1, ba = ab, ca = a^2c, cb = b^2c \rangle.$$

Find the six conjugate classes and the character table of G.

16.15.   Find the conjugate classes and the character table of D_9.

16.16.   Find the conjugate classes and the character table of $S_3 \times Z_3$.

Chapter 17

The Burnside Counting Theorem

In this chapter we'll apply some group theory to counting problems. In particular, a pure group-theoretic result of W. F. Burnside (*Theory of Groups of Finite Order*, 2nd ed. New York: Dover, 1911, reprinted 1955) has been shown to have nice applications to this sort of problem. We'll also see how Burnside's result ties in with characters, as discussed in previous sections.

17.1 Theorem: The Burnside counting theorem

Let G be a group acting on a point set Ω, and for each $g \in G$, let $\chi(g)$ be the number of points of Ω left fixed by g. Then

$$\frac{1}{|G|} \sum_{g \in G} \chi(g)$$

is equal to the number of distinct orbits of Ω under G.

Proof: The action of G on Ω consists of specifying α^g for each $\alpha \in \Omega$ and each $g \in G$, subject to the conditions in Definition 3.1. The idea of this proof is to count the number N of ordered pairs (α, g) for which $\alpha^g = \alpha$, doing this in two different ways. First, fix g. We are then counting the number of points α left fixed by this particular g, which has been denoted in the statement of the theorem by $\chi(g)$. Hence,

$$N = \sum_{g \in G} \chi(g).$$

Now, fix α. Then the number of pairs (α, g) for which $\alpha^g = \alpha$ is given by $|G_\alpha|$, where G_α (as usual) denotes the stabilizer of α under G. Now we'll write Ω as the union of disjoint orbits,

$$\Omega = \alpha_1^G \cup \alpha_2^G \cup \cdots \cup \alpha_r^G,$$

where we have taken r to be the number of distinct orbits of Ω. Then,

$$N = \sum_{\alpha \in \Omega} |G_\alpha| = \sum_{j=1}^{r} \sum_{\alpha \in a_j^G} |G_\alpha|$$

$$= \sum_{j=1}^{r} |\alpha_j^G| |G_{\alpha_j}| \qquad \text{by Proposition 5.10}$$

$$= \sum_{j=1}^{r} |G| \qquad \text{by Theorem 3.13}$$

$$= r \, |G|.$$

Therefore,

$$r = \frac{1}{|G|} \sum_{g \in G} \chi(g),$$

as was to be proved. $\square$

Now let's apply this result to a simple problem.

17.2   Example: A block-coloring problem

Suppose that you are making a set of children's blocks. Each block is a cube, and you will paint the blocks in the set using k different colors. How many distinct colored blocks can you make? For $k = 1$, of course, you have only one color and can make only one color scheme. Try $k = 2$; say, red and blue. Remember that when a child plays with the blocks, they will be moved around according to what you know as the group of rigid symmetries of the cube; thus painting one side blue and the others red can be done in only one way. Here are the possibilities

All six sides red;
One side blue, the other five red;

Two opposite sides blue, the other four red;

Two adjacent sides blue, the other four red;

Three adjacent sides (say, top/front/right) blue, the other three red;

Three "wrapping" sides (say, top/front/bottom) blue, the other three red;

Two adjacent sides red, the other four blue;

Two opposite sides red, the other four blue;

One side red, the other five blue;

All six sides blue.

We have accounted for 10 different colored blocks, merely by listing the possibilities. But that does not seem like a very attractive approach for $k > 2$. In fact, it will turn out that the number of distinct blocks for $k = 3$ is 57, for $k = 4$ is 240, and for $k = 5$ is 800; the chances of enumerating these cases accurately are not good, and it certainly would be laborious. Let's see how the Burnside counting theorem can provide a formula for the number of distinct blocks.

For the set Ω we'll take all colorings of an oriented block; that is, blue on the front and red elsewhere is *different from* blue on the top and red elsewhere. Of course, a rotation of the cube will carry one of these to the other, so they will be in the same orbit of Ω. In fact, the orbits of Ω will be precisely the distinct patterns of colored block that we can paint with k colors. The group G of rigid symmetries of the cube will act on the patterns that are the colorings of an oriented block. Each point of Ω is an assignment of one of k colors to each of the six faces of the cube; hence, $|\Omega| = k^6$. We'll be able to see what the 24 elements of G do by considering the conjugate classes, as in Chapter 15.

C_1 : the identity element

C_2 : six 90° rotations

C_3 : eight 120° rotations

C_4 : six 180° rotations about an axis joining midpoints of edges

C_5 : three 180° rotations about an axis joining midpoints of faces

If e is the identity element, it does not move the cube at all, and so it fixes all k^6 of the points of Ω. An element of C_2 will fix a pattern that has the same color on the four sides being permuted by the rotation (say, front/right/back/left), but we have a free choice for the colors used on the other two faces (say, top/bottom); this leaves us with a total of k^3 patterns left fixed by each of the six rotations in C_2. The 120° rotations in C_3 are each about a main diagonal as axis, so they permute the three faces meeting at one vertex. Those three faces must all have the same color if the pattern is to be unchanged under rotation, so we have k choices of color for three of the faces and another k for the other three, a total of k^2. The 180° rotations about an axis joining the midpoints of two opposite edges permute the six faces in three pairs; hence, these six rotations fix a total of k^3 patterns. Finally, the 180° rotations that are the squares of the rotations in C_2 require that two pairs of opposite faces (say, front/back and left/right) have the same colors, but we again have free choices for the other two faces (top/bottom), making a total of k^4 patterns for each of these three rotations. The values we have found are the 24 $\chi(g)$ referred to in the counting theorem, and the distinct patterns are the orbits. The formula then becomes

$$\frac{1}{|G|} \sum_{g \in G} \chi(g) = \frac{1}{24}(k^6 + 6k^3 + 8k^2 + 6k^3 + 3k^4).$$

We'll summarize this discussion in the following statement.

17.3   Proposition: Distinct patterns for the faces of a cube

If any one of k different colors (or symbols) may be assigned to each face of a cube, then the number of distinct patterns formed (allowing for rotations) is given by

$$\frac{1}{24}(k^6 + 3k^4 + 12k^3 + 8k^2).$$

Now we'll look at some applications of the Burnside counting theorem to questions that could be posed with regard to chemical molecules. The ammonia molecule NH_3 has the form of a trigonal pyramid with an equilateral base and isosceles sides. Zero to three of the hydrogens at the vertices of the base may be replaced by chlorines; how many visually distinct molecules result? The *rigid* symmetries of this pyramid are precisely the rotations of the base; thus the group is cyclic of order 3. The point set Ω now consists of the $2^3 = 8$ *configurations* that one can make by choosing H or Cl at each of the three vertices of the base. A rotation r through $2\pi/3$ will leave fixed only the two configurations having H at all three base positions, or Cl at all three. In terms of Theorem 17.1, we have the count

$$\frac{1}{3}(8 + 2 + 2) = 4$$

orbits, which is to say that the eight configurations reduce to four visually distinct ones. Similarly, if we allow k substituents at each of the H positions,

the result is

$$\frac{1}{3}(k^3 + 2k)$$

orbits.

In the preceding discussion, we treated the pyramid as a rigid solid; the patterns

and

are distinct because the pyramid cannot be inverted. These patterns correspond in a theoretical context to *optical isomers*. For experimental work, inversion of the pyramid (which is geometrically equivalent to turning the solid pyramid inside out) would enable us to transform one isomer into the other and would reduce the number of molecules. If we allow such inversion, we are now dealing with the dihedral group D_3 and the orbits using k substituents would account for

$$\frac{1}{6}(k^3 + 2k + 3k^2)$$

distinct molecules.

The discussion here is an illustration of how group theory can be used to tell what will *not* occur, a characterization expressed by some chemists. Specifically, the argument just concluded tells that from a mathematical point of view, *not more than*

$$\frac{1}{6}(27 + 2(3) + 3(9)) = 10$$

distinct molecules can result from substitution of Cl or of CH_3 for the H in the ammonia molecule. Whether all 10 such distinct molecules exist in nature or can be synthesized is a question for the chemist.

The use of the letter χ in the statement of Theorem 17.1 suggests that a group character is involved and, indeed, one is. Think of the points of Ω as the numbers $1, 2, \ldots, n$ and identify the natural basis vectors $e_1, e_2, \ldots, e_n$ with them, as we did in Chapter 16. Then, as in that discussion, for $g \in G$, take $T(g)$ to be the permutation matrix that carries e_i to e_j if and only if $i^g = j$, and let χ be the character associated with T. Then $\chi(g)$ counts the number of points i of Ω for which $i^g = i$, as in the statement of the theorem. But in the notation of our inner product,

$$\langle \chi, \zeta^{(1)} \rangle = \frac{1}{|G|} \sum_{g \in G} \chi(g),$$

where $\zeta^{(1)}$ is the trivial character of G. Hence, we have proved the following.

17.4   Theorem: The trivial character and orbits

Let G be a group acting on a point set Ω, and for each $g \in G$, let $\chi(g)$ be the number of points of Ω left fixed by g. Then χ is a character of G, and the multiplicity of the trivial character in χ is equal to the number of distinct orbits of Ω under G.

For applications in chemistry, $\zeta^{(1)}$ (which may be denoted, depending upon the group, by A, A^1, A_g, A_1, etc.), is called the *totally symmetric character*, and its multiplicity is of interest. Vibrations of molecules that transform as the totally symmetric character may be observed using Raman spectroscopy.

Exercises

17.1.   Verify the conclusion of the Burnside counting theorem for S_4, regarded as a group acting on the eight vertices of the cube, and for A_4, regarded as a group acting on the four vertices of the tetrahedron.

17.2.   The benzene molecule C_6H_6 may be thought of as a hexagon in the plane with a C at each vertex and an H attached to each C and available for substitution. For example, one of the halogens F, Cl, Br, or I might be substituted for an H. The group of the hexagon is D_6, of order 12. Find a result similar to Proposition 17.3 for assigning k different symbols to each vertex of the hexagon.

17.3.   Find a result similar to Proposition 17.3 for assigning k different symbols to each vertex of the pentagon.

17.4.   Find a result similar to Proposition 17.3 for assigning k different symbols to each vertex of the octagon.

17.5.   The methane molecule CH_4 has the form of a tetrahedron with H at each vertex and C at the center of gravity. If substitution of one of the halogens F, Cl, Br, or I, or retention of the H is allowed at each vertex, how many distinct configurations may be formed? The group is A_4.

17.6.   The pentaborane molecule B_5H_9 may be thought of as a square pyramid, as shown in Figure 17.1. (The hydrogens shown at the sides of the base are in fact below the plane of the square, but for our purposes the figure may be simplified as shown.) Suppose that we allow substitution of a halogen (F, Cl, Br, or I) or retention of the H at each of the four H-positions attached to the vertices of the base, and that we do *not* permit inversion. Show that 165 visually distinct configurations result.

17.7.   Show that if inversion is permitted in Exercise 17.6, the number of distinct configurations drops to 120.

17.8.   Suppose that in Exercise 17.7 we allow substitution of a halogen or retention of the H at each of the nine H-positions. How many distinct configurations may be formed?

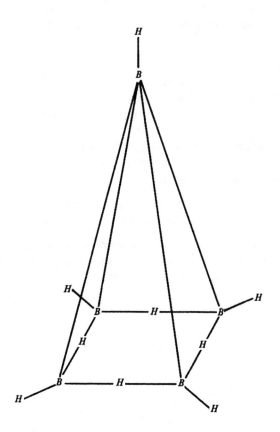

FIGURE 17.1
Pentaborane

Chapter 18

Real Characters

The character table we worked out for D_4 at the end of Chapter 14, as well as those we worked out in Examples 15.9, 15.10, 15.11, and 15.12 contained only real numbers, in fact, integers. You know, from the one-dimensional representations of finite cyclic groups in Proposition 13.1 and of finite abelian groups in 13.6 through 13.12, that complex values may occur in characters. In this chapter we'll see under what circumstances we may or may not expect the characters of a group to take on *only* real values.

Perhaps more interesting, we'll use our results to give two illustrations of how character theory may be used to prove results in abstract group theory, even though the statement of such results never mentions characters.

The first observation is one that you may well have used if you did Exercise 15.10.

18.1 Proposition: Character of an element conjugate to its inverse

Let G be a group and g an element of G that lies in the same conjugate class of G as its own inverse g^{-1}. If χ is any character of G, then $\chi(g)$ is a real number.

Proof: Since, by Proposition 14.6, χ is a class function, and since g and g^{-1} lie in the same conjugate class, $\chi(g) = \chi(g^{-1})$. But by Proposition 14.13, $\chi(g^{-1}) = \overline{\chi(g)}$. Hence, $\chi(g) = \overline{\chi(g)}$, which is to say that $\chi(g)$ is real. $\qquad\square$

Now we'll bring the idea of complex conjugate to bear on an arbitrary representation of a group.

18.2   Definition: Conjugate representation and character

Let G be a finite group, T be a representation of G, and χ be the character associated with T. Then the *conjugate representation of T* is the function defined by

$$T^*(g) = \overline{T(g)} \qquad \text{for each } g \in G.$$

The character associated with T^* is called the *conjugate character of χ* and is denoted χ^*.

That T^* is indeed a representation of G follows quickly from Proposition 14.11.

Considering that g and g^{-1} need not lie, as we assumed they did in Proposition 18.1, in the same conjugate class, but recalling Proposition 14.13, we'll make a corresponding definition for conjugate classes. If G is a finite group and C_j a conjugate class of G, define

$$C_j^* = \{g^{-1} : g \in C_j\}.$$

We call C_j^* the *inverse* of C_j. Certainly, if the elements of C_j have order 2, then $C_j = C_j^*$, and in such examples as S_4 (Proposition 13.12) and D_5 (Exercise 15.10), we had elements of order greater than 3 that were in the same conjugate class as their inverses. This leads to the formulation of the following.

18.3   Definition: Self-inverse conjugate class

Let G be a finite group and C_j a conjugate class of G. If $C_j = C_j^*$, then C_j is called a *self-inverse conjugate class*.

We really should verify that C_j^* is a conjugate class of G. To see this, observe that

$$
\begin{array}{llll}
g, h \in C_j & \text{iff} & x^{-1}gx = h & \text{for some } x \in G \\
& \text{iff} & (x^{-1}gx)^{-1} = h^{-1} & \text{for some } x \in G \\
& \text{iff} & x^{-1}g^{-1}x = h^{-1} & \text{for some } x \in G \\
& \text{iff} & g^{-1}, h^{-1} & \text{are conjugate in } G.
\end{array}
$$

Now we need to check that irreducibility is preserved by the conjugation of a representation and of its character.

18.4 Proposition: Irreducibility of a conjugate character

If $\zeta^{(i)}$ is an irreducible character of a finite group G, then the conjugate character $\zeta^{(i)*}$ is also irreducible.

Proof: Let Z be the representation whose character is $\zeta^{(i)}$; then Z is irreducible. Define $T(g) = (Z(g^{-1}))^t$, where the superscript t denotes, as usual, the transpose of the matrix. Then for $g, h \in G$, we have

$$
\begin{aligned}
T(gh) &= (Z((gh)^{-1}))^t = (Z(h^{-1}g^{-1}))^t \\
&= (Z(h^{-1})Z(g^{-1}))^t = (Z(g^{-1}))^t(Z(h^{-1}))^t \\
&= T(g)T(h),
\end{aligned}
$$

so T is a representation of G. Moreover, if T is reducible, then by Maschke's theorem (Theorem 12.10), T is decomposable, and there is a nonsingular matrix A such that

$$
A^{-1}T(g)A = \begin{bmatrix} T_1(g) & 0 \\ 0 & T_2(g) \end{bmatrix} \qquad \text{for each } g \in G.
$$

But from linear algebra,

$$
A^t(A^{-1})^t = (A^{-1}A)^t = I, \qquad \text{so} \qquad (A^{-1})^t = (A^t)^{-1},
$$

and from this remark we have, for each $g^{-1} \in G$,

$$
\begin{bmatrix} T_1(g)^t & 0 \\ 0 & T_2(g)^t \end{bmatrix} = (A^{-1}T(g)A)^t
$$

$$
= A^t(T(g))^t(A^{-1})^t
$$

$$
= A^t Z(g^{-1})(A^t)^{-1},
$$

so Z is reducible, a contradiction. Hence, T is irreducible. Now if χ is the character of T, then χ is irreducible and

$$\chi(g) = \operatorname{tr}((Z(g^{-1}))^t) = \operatorname{tr}(Z(g^{-1}))$$
$$= \zeta^{(i)}(\vec{g}) = \zeta^{(i)*}(g)$$

for all $g \in G$, as was to be proved.

Since the columns of a character table are mutually orthogonal in the sense of Corollary 15.7, the table itself forms a nonsingular matrix. We'll use this observation in the results that follow.

18.5   Lemma: Fixed points of two permutations

Let A be a nonsingular $n \times n$ matrix and let π and ϕ be permutations of $\{1,\ldots,n\}$; thus, $\pi,\phi \in S_n$. Let A_1 be the matrix whose ith row is the $\pi(i)$th row of A and A_2 be the matrix whose jth column is the $\phi(j)$th column of A. If $A_1 = A_2$, then π and ϕ leave fixed the same number of points.

Before we prove this lemma, let's examine what it describes. Exercises 18.1 through 18.3 provide both an example in which π and ϕ do indeed leave fixed the same number of points and a counterexample to the converse of the lemma. The example here will be one in which π and ϕ do *not* leave fixed the same number of points, but we have chosen it for visual clarity as to the row and column switches described in the lemma. Take $\pi = (13)$ and $\phi = (132)$, both in S_3, and let

$$A = \begin{bmatrix} 1 & 2 & 3 \\ 4 & 5 & 6 \\ 7 & 8 & 0 \end{bmatrix}.$$

Then the first row of A_1 is the third row of A, and the third row of A_1 is the first row of A. The second row of A_1 is the same as the second row of A because π does not move 2. Thus,

$$A_1 = \begin{bmatrix} 7 & 8 & 0 \\ 4 & 5 & 6 \\ 1 & 2 & 3 \end{bmatrix}.$$

By similar reasoning,

$$A_2 = \begin{bmatrix} 3 & 1 & 2 \\ 6 & 4 & 5 \\ 0 & 7 & 8 \end{bmatrix}.$$

Of course, this example does not have $A_1 = A_2$, nor do π and ϕ fix the same number of points, but you should go through it carefully to see how A_1 and A_2 are constructed from the matrix A. Now we'll turn to the proof of Lemma 18.5.

Proof: Represent π and ϕ by permutation matrices in the usual way, that is, represent π by B and ϕ by C, where

$$Be_{\pi(i)} = e_i \qquad \text{and} \qquad Ce_i = e_{\phi(i)}$$

for the column vectors $e_1, \ldots, e_n$ forming the natural basis of $\mathbf{R}^n$. The result is that $B = (b_{ij})$, where $b_{ij} = 1$ if $i = \pi(j)$ and $b_{ij} = 0$ otherwise. Using the Kronecker delta (see Chapter 15), we can express this observation as

$$B = (\delta_{j\pi(j)});$$

similarly,

$$C = (\delta_{\phi(i)i}).$$

It follows that $A_1 = BA$ and $A_2 = AC$. But $A_1 = A_2$, so $BA = AC$, and hence $A^{-1}BA = C$. Hence by Proposition 14.4, $\text{tr}(B) = \text{tr}(C)$. But by the formation of B and C (observing that $b_{ii} = 1$ if and only if $\pi(i) = i$, and similarly for c_{ii}), the trace of B is the number of points left fixed by π, and the trace of C is the number of points left fixed by ϕ. This completes the proof. $\square$

This lemma allows us to prove the first of two theorems about real characters.

18.6 Theorem: Real irreducible characters of a finite group

The number of *real* irreducible characters of a finite group G equals the number of self-inverse conjugate classes of G.

Proof: Let s denote (as usual) the number of distinct conjugate classes of G. By Proposition 18.4, the mapping $\zeta^{(i)} \to \zeta^{(i)*}$ is a permutation of the irreducible characters of G and hence of the integers $1, 2, \ldots, s$; let it be the π of Lemma 18.5. From the discussion of the inverse of a conjugate class C_j, the mapping $C_j \rightarrow C_j^*$ is a permutation of the conjugate classes (and likewise of $1, 2, \ldots, s$); let it be the ϕ of the lemma. Now think of the character table of G as a matrix A whose (i, j)-entry is $\zeta_j^{(i)}$. If we apply π to A as described in the lemma, we obtain

$$A_1 = (\overline{\zeta_j^{(i)}}),$$

and if we apply ϕ to A as in the lemma, we have

$$A_2 = (\zeta_j^{(i)*}).$$

By Proposition 14.13,

$$\begin{aligned} \zeta_j^{(i)*} &= \zeta^{(i)}(x^{-1}) \qquad \text{for any } x \in C_j \\ &= \overline{\zeta^{(i)}(x)} \\ &= \overline{\zeta_j^{(i)}}; \end{aligned}$$

hence, $A_1 = A_2$. Therefore, by Lemma 18.5, π and ϕ leave fixed the same number of points of the set $\{1, \ldots, s\}$. But π leaves fixed precisely the real characters, and ϕ fixes precisely the self-inverse conjugate classes. This completes the proof.   □

Now in an abelian group, every *element* is self-conjugate, so each conjugate class consists of one element alone. Hence, a conjugate class is self-inverse if and only if the element comprising it has order 1 or 2. Thus, we have proved the following.

18.7   Corollary: Real characters of a finite abelian group

The number of real irreducible characters of a finite abelian group G is equal to the number of elements of order 1 or 2 in that group.

Then as a consequence of Theorem 18.6, we have our other result on real characters as follows.

18.8 Theorem: A group of odd order has only one real character

If $|G|$ is odd, then G has only *one* real irreducible character, namely, the trivial character $\zeta^{(1)}$, which sets $\zeta^{(1)}(g) = 1$ for all $g \in G$.

Proof: Let C_j be a conjugate class for which $C_j = C_j^*$ and let $x \in C_j$. Then there is some $u \in G$ for which $u^{-1}xu = x^{-1}$. Hence,

$$u^{-2}xu^2 = u^{-1}(u^{-1}xu)u = u^{-1}x^{-1}u = (u^{-1}xu)^{-1} = x,$$
$$u^{-3}xu^3 = u^{-1}(u^{-2}xu^2)u = u^{-1}xu = x^{-1},$$

and similarly,

$$u^{-r}xu^r = x \qquad \text{when } r \text{ is even,}$$
$$u^{-r}xu^r = x^{-1} \qquad \text{when } r \text{ is odd.}$$

Now by Proposition 2.11 and the theorem of Lagrange (Theorem 3.10), the order r of u divides the order of G, which is odd. Suppose, then, that r is the order of u; we have

$$x = 1 \cdot x \cdot 1 = u^{-r}xu^r = x^{-1}.$$

But an element that equals its own inverse must have order 1 or 2, and since $|G|$ is odd, x cannot have order 2. Therefore, $x = 1$, and the only self-inverse conjugate class is $C_1 = \{1\}$. Hence by Theorem 18.6, G has only one irreducible character. But every group has the trivial character $\zeta^{(1)}$, which assigns 1 to every group element. This completes the proof. $\square$

An easy illustration of Theorem 18.8 is given by the cyclic group of order 3 which, by Proposition 13.1 and Theorem 13.12, has the trivial character and two complex ones (each involving a primitive cube root of 3 and its square). A similar observation holds, of course, for any cyclic group of odd order.

However, let's *apply* Theorem 18.8 rather than merely illustrating it. All of the odd numbers from 3 through 19 are prime except 9 and 15; by Proposition 13.11 the two groups of order 9 are abelian, and by Exercise 7.1 the only group of order 15 is cyclic. Hence, the smallest possibility for an example is a group of order 21. By Theorem 13.8 the only abelian group of order 21 is cyclic, but indeed there is exactly one nonabelian group of this order, and we'll use this as our example.

18.9   Example: The nonabelian group of order 21

The only nonabelian group of order 21 is given by

$$G = \langle x, y : x^3 = y^7 = 1, yx = xy^2 \rangle.$$

If we set out to construct the character table (as Exercise 18.6 asks you to do), we know to look for complex values in every character *except* $\zeta^{(1)}$. Alternatively, you might say that Theorem 18.8 serves as a check in computing the character table since, if you find only real values for any character other than $\zeta^{(1)}$, you know you have made an error, and you might as well stop there and retrace your steps.

As promised, we'll conclude this chapter with two results in abstract group theory which can be proved using group character theory. To give these proofs, we'll need to invoke the following powerful theorem, which is not at all easy to prove, but which you can readily apply, as we have done with such results as the Sylow theorems, the orthogonality relations, and the main theorem on finite abelian groups.

18.10   Theorem: Degree of irreducible character divides order of group

If $\zeta^{(i)}$ is an irreducible character of a finite group G, then the degree of $\zeta^{(i)}$ divides the order of G.

The proposition that follows does not *require* a character-theoretic proof; indeed, Exercise 18.7 asks you to give a proof using more elementary methods. The point of this particular illustration is to see that character theory can be used to prove results in abstract group theory.

18.11   Proposition: On some groups of order pq

If p and q are primes with $p > q$, and if q does not divide $p - 1$, then the only group of order pq is cyclic.

Character-Theoretic Proof: First, note that if we allowed $p = q$, then G would have order p^2, and G would be abelian by Proposition 13.11, but not necessarily cyclic.

Let $z_1, \ldots, z_s$ be (as is our custom) the degrees of the irreducible characters of G, listed so that $z_i \leq z_{i+1}$ for $1 \leq i \leq s - 1$. By Corollary 15.8,

$$|G| = pq = z_1^2 + \cdots + z_s^2.$$

If $[G : G'] = r$, then $z_i = 1$ for $1 \leq i \leq r$ because we arranged the irreducible characters in nondecreasing order of degree, and $z_i > 1$ for $r + 1 \leq i \leq s$. Consider some i for which $z_i > 1$ (and $i > r$); by Theorem 18.10, z_i divides pq, and since pq is the product of two primes, z_i must be equal to one of p, q, or pq. If $z_i = p$, then since $p > q$,

$$z_i^2 = p^2 > pq = z_1^2 + \cdots + z_s^2 > z_i^2,$$

which is a contradiction. Similarly, $z_i \neq pq$. Hence, $z_i = q$ for all $i > r$. Now

$$pq = z_1^2 + \cdots + z_r^2 + z_{r+1}^2 + \cdots + z_s^2$$
$$= 1(r) + q^2(s - r).$$

Since q divides the other terms in this equation, q must divide r also. But $r = [G : G']$, which in turn is a divisor of $|G|$; hence r divides $|G|$, and so r must equal q or pq. If $r = q$, then the preceding equation factors as

$$pq = q + q^2(s - r),$$
$$p = 1 + q(s - r),$$
$$p - 1 = q(s - r),$$

and since $s - r$ is an integer, q divides $p - 1$, a contradiction to the hypothesis. Hence $pq = r = [G : G']$, and $|G'| = 1$; that is, G is abelian. But then, by Theorem 13.8ii, G is cyclic, as was claimed. □

We'll conclude this chapter with a striking result due to Burnside, the proof of which is an elegant example of the use of character theory as a proof technique. We'll make use of the congruence notation introduced in connection with Theorem 7.13.

182   GROUPS AND CHARACTERS

18.12 Theorem: Conjugate classes of a group of odd order

Let G be a group of odd order having s conjugate classes, then $s \equiv |G|$ mod 16.

Proof: By Theorem 18.8 the only real irreducible character of G is the trivial character $\zeta^{(1)}$; by Proposition 18.4 the remaining irreducible characters come in (distinct) conjugate pairs. Hence the number of conjugate classes (which equals the number of irreducible characters) is odd; let's write $s = 2m - 1$, where m is an integer and then list the irreducible characters of G as

$$\zeta^{(1)}, \zeta^{(2)}, \zeta^{(2)*}, \ldots, \zeta^{(m)}, \zeta^{(m)*}.$$

Now by Theorem 18.10, if

$$z_i = \zeta_1^{(i)} = \zeta_1^{(i)*},$$

with $1 < i \leq m$, then z_i is odd. By Corollary 15.8,

$$|G| = 1 + z_2^2 + z_2^2 + \cdots + z_m^2 + z_m^2$$

$$= 1 + \sum_{i=2}^{m} 2z_i^2,$$

and since z_i^2 is odd, we can write each $z_i = 2y_i + 1$, where $1 < i \leq m$ and each y_i is a nonnegative integer. Hence,

$$|G| = 1 + \sum_{i=2}^{m} 2(2y_i + 1)^2$$

$$= 1 + 2\sum_{i=2}^{m}(4y_i^2 + 4y_i + 1)$$

$$= 1 + 2(m - 1) + 2\sum_{i=2}^{m} 4y_i(y_i + 1)$$

$$= 2m - 1 + 8\sum_{i=2}^{m} y_i(y_i + 1).$$

Now either y_i or $y_i + 1$ must be even for each i, so we may replace each term $y_i(y_i + 1)$ in the summation with $2x_i$, where x_i is again a nonnegative integer. Then since we've set $s = 2m - 1$, we can conclude the proof that $s \equiv |G| \bmod 16$ as follows:

$$|G| = s + 8 \sum_{i=2}^{m} 2x_i$$

$$= s + 16 \sum_{i=2}^{m} x_i$$

where

$$\sum_{i=2}^{m} x_i$$

is, as required, a nonnegative integer. If G is abelian, then of course $s = |G|$, and all $x_i = 0$. □

A significant feature of group character theory is that it affords proofs of many such abstract group-theoretic theorems; such results may be found in the references listed in the Bibliography.

Exercises

18.1. As a further illustration of how the matrices A_1 and A_2 are formed in Lemma 18.5, consider

$$A = \begin{bmatrix} 1 & 2 & 0 & 3 \\ 1 & 3 & 2 & 0 \\ 1 & 0 & 3 & 2 \\ 4 & 5 & 5 & 5 \end{bmatrix}$$

with the permutations $\pi = (132)$ and $\phi = (234)$, which (unlike the example given in the text) do fix the same number of points. Find A_1 and A_2, and verify that $A_1 = A_2$.

18.2. For the matrix A of Exercise 18.1, form the matrices B and C from the proof of Lemma 16.5. Verify that B and C have the same number of entries of 1 on the main diagonal and that $A_1 = BA = AC = A_2$.

18.3. Let
$$A = \begin{bmatrix} 1 & 2 & 3 \\ 4 & 5 & 6 \\ 7 & 8 & 0 \end{bmatrix},$$

$\pi = (13)$, and $\phi = (23)$. Verify by this example that the *converse* of Lemma 18.5 is false.

18.4. Proposition 18.1 shows that if g is an element of order 2 in a group G and if χ is any character of G, then $\chi(g)$ must be real. Show that, in fact, $\chi(g)$ must be an integer.

18.5. Define a group H by

$$H = \langle u, v : u^3 = v^7 = 1, vu = uv^4 \rangle.$$

First, prove that H has order 21. Then show that H is isomorphic to the group G of Example 18.9.

18.6. How many conjugate classes will the group G of Example 18.9 have? Find these conjugate classes and the character table of G.

18.7. Use the Sylow theorems to produce an alternative proof of Proposition 18.11.

18.8. Let p and q be primes with $p > q$, and assume that q does not divide $p^2 - 1$. Use the Sylow theory as in Exercise 18.7 to prove that a group of order $p^2 q$ must be abelian. If such a group is abelian, must it be cyclic, as in Proposition 18.11?

Chapter 19

Induced Characters

In Chapter 13, we used representations of a factor group G/N to find representations of the group G. It is also possible, although a bit more complicated, to lift a representation of a subgroup H of G to a representation of G itself. If you reflect, you'll realize that this second construction is naturally more difficult: instead of "burying" a bit of the given group G in a kernel N, we look at a subgroup H which is "surrounded" by *Terra Incognita* in the group G. Yet it turns out that this construction is a useful and powerful tool for finding new representations, in particular those that are irreducible. As a result, we'll be able to complete the project, begun at the end of Chapter 15, of constructing the character table of the alternating group A_5 of order 60. *All groups in Chapter 19 will be assumed to be finite.*

19.1 Definition: Induced representation and character

Let $H \leq G$, and let T be a representation of H afforded either by $\mathbf{R}^m$ or by $\mathbf{C}^m$. Define a function $\dot{T}$ on G by

$$\begin{cases} \dot{T}(g) = T(g) & \text{if } g \in H, \\ \dot{T}(g) = \mathbf{0}_m & \text{if } g \notin H, \end{cases}$$

where m is the degree of T and $\mathbf{0}_m$ denotes the $m \times m$ matrix consisting entirely of zeros. Now let $n = [G : H]$ and write

$$G = H \cup g_2 H \cup g_3 H \cup \cdots \cup g_n H,$$

taking the identity 1 of G as the representative g_1 of the first coset. Now for $g \in G$, we define each $T^G(g)$ to be an $mn \times mn$ matrix consisting of n^2 submatrices, each $m \times m$, arranged in n rows and n columns, with the

(i, j)-subblock given by

$$T^G(g) = (\dot{T}(g_i^{-1}gg_j)).$$

Then T^G is called an *induced representation* of G, or more specifically, the representation of G induced by the representation T of H. If χ denotes the character associated with T, then we denote the character of T^G by χ^G and call χ^G an *induced character*.

Since the degree of T in the example above was m and the index of H in G was n, the induced representation $T^G(g)$ has degree mn. Each of the n^2 submatrices is equal either to some $T(h)$ or to $\mathbf{0}_m$.

The proof that T^G is indeed a representation of G is something of a notational *tour de force* (left to Exercise 19.7); the essence of the argument is to show that the submatrices of $T^G(g)$ having the form $T(h)$ occur exactly once in each of the n rows and exactly once in each of the n columns, so that $T^G(g)$ resembles a permutation matrix, but with nonzero blocks in place of 1 and $\mathbf{0}_m$ in place of 0.

Let's look at a couple of examples, using your old friend D_4:

$$G = \langle r, c : r^4 = c^2 = 1, cr = r^{-1}c \rangle,$$
$$H = \langle r \rangle, \qquad G = H \cup cH.$$

Let $T(r) = [i]$ be a representation of H; then

$$\dot{T}(1) = [1], \qquad \dot{T}(r) = [i], \qquad \dot{T}(r^2) = [-1], \qquad \dot{T}(r^3) = [-i],$$
$$\dot{T}(c) = \dot{T}(rc) = \dot{T}(r^2c) = \dot{T}(r^3c) = [0].$$

Here $m = 1$, $n = 2$, $g_1 = 1$, and $g_2 = c$. When $g = r$, the elements $g_i^{-1}gg_j$ are, respectively,

$$\begin{array}{ll} 1\,r\,1 = \ \ r & 1\,r\,c = rc \\ c\,r\,1 = r^3c & c\,r\,c = r^3. \end{array}$$

Definition 19.1 then gives

$$T^G(r) = \begin{bmatrix} i & 0 \\ 0 & -i \end{bmatrix}.$$

When $g = c$, the elements $g_i^{-1}gg_j$ are

$$\begin{array}{ll} 1\,c\,1 = \ \ c & 1\,c\,c = 1 \\ c\,c\,1 = 1c & c\,c\,c = c, \end{array}$$

from which you can calculate

$$T^G(c) = \begin{bmatrix} 0 & 1 \\ 1 & 0 \end{bmatrix}.$$

You can easily show that the associated character χ^* is given by

$$\chi^*(1) = 2, \qquad \chi^*(r^2) = -2, \qquad \chi^*(r) = 0, \qquad \chi^*(c) = 0, \qquad \chi^*(rc) = 0.$$

Hence, $\chi^* = \zeta^{(5)}$ from the character table for D_4 found at the end of Chapter 14.

What you should notice in this example is that an irreducible matrix representation for D_4 was constructed directly as an induced representation and that irreducibility could be verified from the character table (see also Exercise 19.1). In general, an induced representation need not be irreducible, but once such a representation is constructed, we can use Theorem 16.1 to reduce the associated character to its irreducible components. After the next example, we'll work out induced characters without even referring to the matrix representations that correspond to the characters of the subgroup.

The second example is again from D_4 but now $m = n = 2$. Consider

$$T(r) = \begin{bmatrix} 0 & -1 \\ 1 & 0 \end{bmatrix};$$

then, taking matrix powers, we have

$$T(r^2) = \begin{bmatrix} -1 & 0 \\ 0 & -1 \end{bmatrix} \qquad \text{and} \qquad T(r^3) = \begin{bmatrix} 0 & 1 \\ -1 & 0 \end{bmatrix}.$$

Definition 19.1 then gives

$$T^G(r) = \begin{bmatrix} 0 & -1 & 0 & 0 \\ 1 & 0 & 0 & 0 \\ 0 & 0 & 0 & 1 \\ 0 & 0 & -1 & 0 \end{bmatrix}, \qquad T^G(c) = \begin{bmatrix} 0 & 0 & 1 & 0 \\ 0 & 0 & 0 & 1 \\ 1 & 0 & 0 & 0 \\ 0 & 1 & 0 & 0 \end{bmatrix}.$$

Reduction of the associated character χ^* shows that $\chi^* = 2\zeta^{(5)}$.

Now, as promised, we'll see how to construct an induced character directly from a character of a subgroup. Let H be a subgroup of a group G and χ be a character of H. Define $\dot\chi$ by analogy with T as

$$\begin{cases} \dot\chi(g) = \chi(g) & \text{if } g \in H, \\ \dot\chi(g) = 0 & \text{if } g \notin H. \end{cases}$$

Once again writing

$$G = H \cup g_2 H \cup g_3 H \cup \cdots \cup g_n H,$$

we can set

$$\chi^G(g) = \sum_{i=1}^{n} \dot\chi(g_i^{-1} g g_i)$$

with each $\chi^G(g)$ being, clearly, the trace of the induced $T^G(g)$. We have proved the following.

19.2   Proposition: Formation of an induced character

Let $H \leq G$, $\{1, g_2, g_3, \ldots, g_n\}$ be a set of left coset representatives for H in G, and χ be a character of H. Then the induced character on G is given by

$$\chi^G(g) = \sum_{i=1}^{n} \dot\chi(g_i^{-1} g g_i).$$

Now back in Chapter 15, we had two ways of writing the inner product $\langle \chi, \psi \rangle$, the one in Definition 15.1, in which the sum was taken over all elements of G, and the one in Proposition 15.3, in which the sum was taken over the conjugate classes of G. The former is usually appropriate when one is trying to prove a theorem, while the latter may be preferable for computations. With induced characters we again have two forms, the one above (involving n terms in the summation) more useful in calculations, and another (in which the coset representatives do not appear) more suitable for proofs.

19.3   Proposition: A characterization of the induced character

Let $H \leq G$ and let χ be a character of H. Then the character of G induced from χ is given by

$$\chi^G(g) = \frac{1}{|H|} \sum_{x \in G} \dot\chi(x^{-1} g x).$$

Proof: If $g \in G$ and $h \in H$, then $h^{-1}gh$ is a conjugate of g and so

$$\dot{\chi}(h^{-1}gh) = \dot{\chi}(g)$$

by Proposition 14.6. (Note that $h^{-1}gh \in H$ if and only if $g \in H$.) Hence for each i and for each $h \in H$,

$$\dot{\chi}((g_i h)^{-1} g(g_i h)) = \dot{\chi}(g_i^{-1} g g_i);$$

consequently, for $i = 1, 2, \ldots, s$,

$$\chi(g_i^{-1} g g_i) = \frac{1}{|H|} \sum_{x \in g_i H} \dot{\chi}(x^{-1} g x).$$

Substituting these terms into the expression in Proposition 19.2 gives the desired equation. $\qquad\square$

Let's see how Proposition 19.3 can be used in proving a result about induced characters.

19.4 Proposition: Character induced from a normal subgroup

Let $H \trianglelefteq G$ and let χ be a character of H. Then $\chi^G(g) = 0$ for all elements $g \notin H$.

Proof: For any $x \in G$, conjugation by x leaves H invariant and moves no element of G that is outside H into H. Hence in Proposition 19.3, all of the $\dot{\chi}(x^{-1}gx)$ in the summation are equal to 0 when $g \notin H$. $\qquad\square$

This observation reduces the labor of computing the induced character when H is a normal subgroup of G.

Proposition 19.3 also removes an objection that might be raised to the formulation of the induced representation in Definition 19.1, namely, the fact that one particular choice of left coset representatives appears in the defining equations. However, the sum in Proposition 19.3 is taken over all elements of G, and no such choice is in evidence. This establishes the following.

19.5   Proposition: Induced character is unambiguously defined

For a given subgroup H of a group G and a given character χ of H, the values of the induced character χ^G are independent of the choice of coset representatives in the definition of the induced representation.

In Proposition 13.16 we restricted the domain of a one-dimensional representation to the subgroup G' of G; similarly, a character may likewise have its domain restricted to a subgroup of the group on which it has been defined. If G is a group, T a representation of G, and H a subgroup of G, we'll denote by $T\big|_H$ the restriction of T to H, specifically, $T\big|_H(h) = T(h)$ if $h \in H$, and $T\big|_H(g)$ is undefined when $g \notin H$. If χ is a character of G, $\chi\big|_H$ will similarly denote the restriction of χ to H.

Restricting the domain of a character to a subgroup is, in a sense, the reverse of inducing a character of a group from one of a subgroup, and we'll see now that these two procedures are related through the fact that inner products can be taken in either H or G.

19.6   Theorem: Inner products of restricted and induced characters

If $H \leq G$, if χ is a character of G, and if μ is a character of H, then

$$\langle \mu, \chi\big|_H \rangle = \langle \mu^G, \chi \rangle.$$

Proof: First,

$$\langle \mu^G, \chi \rangle = \frac{1}{|G|} \sum_{g \in G} \mu^G(g) \overline{\chi(g)}$$

$$= \frac{1}{|G|} \frac{1}{|H|} \sum_{g \in G} \sum_{x \in G} \dot{\mu}(x^{-1}gx) \overline{\chi(g)} \qquad \text{by Proposition 19.3}$$

$$= \frac{1}{|G||H|} \sum_{g \in G} \sum_{x \in G} \dot{\mu}(x^{-1}gx) \overline{\chi(x^{-1}gx)} \qquad \text{by Proposition 14.6.}$$

Now since conjugation by x is an automorphism of G, a summation over all $g \in G$ is precisely the same as a summation over all $x^{-1}gx \in G$. Hence, set $y = x^{-1}gx$; then

$$\langle \mu^G, \chi \rangle = \frac{1}{|G||H|} \sum_{y \in G} \sum_{x \in G} \dot{\mu}(y) \overline{\chi(y)}.$$

But now x does not appear in the summands, and hence the effect of the summation over $x \in G$ is merely to add up the same term $|G|$ number of times. Hence,

$$\langle \mu^G, \chi \rangle = \frac{1}{|H|} \sum_{y \in G} \dot{\mu}(y) \overline{\chi(y)}$$

$$= \frac{1}{|H|} \sum_{y \in H} \mu(y) \overline{\chi(y)}$$

$$= \langle \mu, \chi|_H \rangle$$

since $\dot{\mu}(y) = 0$ whenever $y \notin H$. This completes the proof.

As a consequence of Theorem 19.6 we have the following famous result.

19.7 Theorem: Frobenius reciprocity theorem

Let $H \leq G$, let χ be an irreducible character of G, and let μ be an irreducible character of H. Then the multiplicity of χ in μ^G is equal to the multiplicity of μ in $\chi|_H$.

Proof: In the notation of Theorem 16.1, what we have to prove is that

$$\langle \mu^G, \chi \rangle = \langle \chi|_H, \mu \rangle.$$

Now $\langle \chi|_H, \mu \rangle$ is a nonnegative (real) integer, so

$$\langle \chi|_H, \mu \rangle = \overline{\langle \chi|_H, \mu \rangle} = \langle \mu, \chi|_H \rangle$$

by Proposition 15.2. The result now follows immediately by Theorem 19.6. $\quad\square$

To illustrate the Frobenius reciprocity theorem, consider the irreducible characters of A_4 and of S_4 that we found in Examples 15.11 and 15.12. Write $G = S_4$, $H = A_4$; let $\chi = \zeta^{(4)}$ from the character table for S_4, and $\mu = \zeta^{(4)}$ from the character table for A_4. Then the values of $\chi|_H$ may be tabulated as

$$\chi|_H \quad \begin{array}{|cccc} 3 & -1 & 0 & 0 \end{array}$$

and clearly $\chi|_H = \mu$, so we have $\langle \chi|_H, \mu \rangle = 1$. Hence, $\langle \mu^G, \chi \rangle = 1$ also. Now the equality $\chi|_H = \mu$ makes sense because $\deg(\chi|_H) = \deg(\mu) = 3$, but since $\deg(\mu^G) = 6$, all we know thus far is that $\mu^G = \chi + \psi = \zeta^{(4)} + \psi$, where ψ is some (possibly reducible) character of degree 3. Now, using the same G, H, and μ, let $\chi = \zeta^{(5)}$ for S_4, Then,

$$\chi|_H \quad \begin{array}{|cccc} 3 & -1 & 0 & 0 \end{array}$$

and again $\langle \mu^G, \chi \rangle = 1$. Thus, $\mu^G = \zeta^{(4)} + \zeta^{(5)}$; note that both sides now have degree 6. Notice, in addition, that two distinct irreducible characters for S_4 give the same irreducible character of A_4 upon restriction to A_4 and that *both* of these irreducible characters are constituents of the character for S_4 induced from the one irreducible character of A_4.

The power of the Frobenius reciprocity theorem is evident in that we were able to find the values of μ^G and to resolve μ^G into its irreducible components without ever computing the induced representation whose character is μ^G. Of course, now that we know that $\mu^G = \zeta^{(4)} + \zeta^{(5)}$, we can immediately write down the tabulation

$$\mu^G \quad \begin{array}{|ccccc} 6 & 0 & 0 & 0 & -2 \end{array}$$

This looks somewhat like the character ρ of the regular representation of S_4 except for the 6 in place of $\rho(1) = 24$ and -2 for 0. Well, recall from Proposition 16.3 that

$$\rho = \sum_{i=1}^{5} z_i \zeta^{(i)},$$

where $z_i = \deg(\zeta^{(i)})$, so in fact, μ^G is the sum of two of the constituents of ρ.

As a further application of Theorem 19.7, we'll prove two more results.

19.8   Theorem: An upper bound on degrees of irreducible characters

Let $H \leq G$ and ζ be an irreducible character of G. Then there is an irreducible character χ of H such that

$$[G : H] \cdot \deg(\chi) \geq \deg(\zeta).$$

Proof: For the given ζ, let χ be an irreducible character of H that is a constituent of $\zeta|_H$; we have $\langle \chi|_H, \chi \rangle \neq 0$. Then by Theorem 19.7, $\langle \chi^G, \zeta \rangle \neq 0$. Hence,

$$\deg(\zeta) \leq \deg(\chi^G) = [G : H] \cdot \deg(\chi).   \qquad \square$$

19.9   Corollary: Other upper bounds on degrees of irreducible characters

If G has an abelian subgroup H with $[G : H] = n$, then the degree of every irreducible character of G is less than or equal to n. If the maximum of the orders of the elements of G is m, then the degree of every irreducible character of G is less than or equal to $|G|/m$.

Proof: For the first part, since H is abelian, the $\deg(\chi)$ in Theorem 19.8 must be 1. For the second claim, if g is an element of maximum order in G, then $\langle g \rangle$ is an abelian subgroup of G, and the first part applies. $\square$

The question of transitivity is an important one in mathematics. You know that if $K \leq H$ and $H \leq G$, then $K \leq G$ also; on the other hand, $K \trianglelefteq H$ and $H \trianglelefteq G$ does not imply that $K \trianglelefteq G$ (examples abound, the easiest ones being in D_4 and A_4). The procedure for inducing characters of a subgroup, fortunately, follows the positive route: if a character χ is induced first from K to H and thence to G, the result is the same function as one obtains by inducing χ directly from K to G.

19.10   Theorem: Transitivity of induction

Let $K \leq H \leq G$ and let χ be a character of K. Then $(\chi^H)^G = \chi^G$.

Proof: The proof is left as Exercise 19.6. □

To conclude this chapter, we'll complete the character table for A_5, which we stated in Chapter 15 and furthered in Example 16.8.

19.11 Example: The character table for A_5

Where matters at present stand is

	C_1	C_2	C_3	C_4	C_5
h_j	1	15	20	12	12
$\zeta^{(1)}$	1	1	1	1	1
$\zeta^{(2)}$	3				
$\zeta^{(3)}$	3				
$\zeta^{(4)}$	4	0	1	-1	-1
$\zeta^{(5)}$	5				

where C_2 consists of elements of the form $(\alpha\beta)(\gamma\delta)$, C_3 consists of those of the form $(\alpha\beta\gamma)$, C_4 consists of conjugates of (12345), and C_5 consists of conjugates of (12354). Writing $G = A_5$ and $H = A_4$ (where we might, to be specific, take H to be the stabilizer of the point 5), uninteresting calculations lead to the decomposition

$$G = H \cup (152)H \cup (253)H \cup (245)H \cup (354)H.$$

Now let $\chi = \zeta^{(2)}$ from the character table for A_4 (Example 15.11). We compute, for C_1 and C_2,

$$\chi^G(1) = \chi(1) \cdot 5 = 5,$$

$$\chi^G((12)(35)) = \dot{\chi}((12)(35)) + \dot{\chi}((125)(12)(35)(152))$$

$$+ \dot{\chi}((235)(12)(35)(253)) + \dot{\chi}((254)(12)(35)(245))$$

$$+ \dot{\chi}((345)(12)(35)(354))$$

$$= 0 + \dot{\chi}((15)(23)) + \dot{\chi}((15)(23))$$

$$+ \dot{\chi}((14)(23)) + \dot{\chi}((12)(45))$$

$$= 0 + 0 + 0 + 1 + 0$$

$$= 1.$$

Similarly, for C_3, C_4, and C_5,

$$\chi^G((123)) = \zeta + \dot{\chi}((135)) + \dot{\chi}((152)) + \dot{\chi}((143)) + \dot{\chi}((125))$$

$$= \zeta + 0 + 0 + \zeta^2 + 0$$

$$= -1,$$

$$\chi^G((12345)) = 0 + \dot{\chi}((13425)) + \dot{\chi}((15243))$$

$$+ \dot{\chi}((14352)) + \dot{\chi}((12534))$$

$$= 0,$$

$$\chi^G((12354)) = 0.$$

Now $\langle \chi^G, \zeta^{(1)} \rangle = \frac{1}{60}(5 + 15 - 20) = 0$; hence, $\zeta^{(1)}$ is not a constituent of χ^G. Since the only other irreducible characters of A_5 have degrees 3, 3, 4, and 5, it follows that χ^G must be the irreducible character $\zeta^{(5)}$ of degree 5, adding to the table above the fifth row

$\zeta^{(5)}$	5	1	-1	0	0

Let's substitute variables for the unknown values in the second and third rows:

$\zeta^{(2)}$	3	a	c	p	r
$\zeta^{(3)}$	3	b	d	q	s

Apply Corollary 15.7 to the second through fifth columns, each in a product with the first; the result is

$$a + b = -2, \qquad c + d = 0, \qquad p + q = 1, \qquad r + s = 1.$$

Since C_2 and C_3 consist of all elements of order 2 and 3, respectively, they are self-inverse conjugate classes. Moreover, since

$$[(14)(23)]^{-1}(12345)[(14)(23)] = (15432) = (12345)^{-1},$$

C_4 is self-inverse; hence, so is C_5. By Theorem 18.6, all characters of A_5 are real. (This means that we can ignore the complex conjugates in taking inner products.) Apply Corollary 15.7 to the second column (with itself); then,

$$1 + a^2 + b^2 + 1 = 4,$$

writing a^2 rather than $a\bar{a}$ because we know that all entries in the table are real. This equation, together with $a + b = -2$, reduces to $a = -1$, $b = -1$. The same procedure applied to the third column gives $c^2 + d^2 = 0$, whose unique *real* solution is $c = d = 0$. Corollary 15.5 applied to the third and fourth rows gives $r = 1 - p$, and the product of the fourth column with itself yields

$$1 + p^2 + (1 - p)^2 + 1 = \frac{60}{12} = 5,$$

which may be solved by the quadratic formula for

$$p = \frac{1 \pm \sqrt{5}}{2}.$$

Now $1 - p$ is the same as p except for the opposite choice of sign, so we may as well take $+$ in p and $-$ in $1 - p$. This finally completes the following character table for A_5:

	C_1	C_2	C_3	C_4	C_5
h_j	1	15	20	12	12
$\zeta^{(1)}$	1	1	1	1	1
$\zeta^{(2)}$	3	-1	0	$\frac{1+\sqrt{5}}{2}$	$\frac{1-\sqrt{5}}{2}$
$\zeta^{(3)}$	3	-1	0	$\frac{1-\sqrt{5}}{2}$	$\frac{1+\sqrt{5}}{2}$
$\zeta^{(4)}$	4	0	1	-1	-1
$\zeta^{(5)}$	5	1	-1	0	0

Exercises

19.1.   Consider the representation T of A_4 given by

$$T((123)) = [\zeta], \qquad T((124)) = [\zeta^2],$$

where ζ is a primitive cube root of 1. (Note that T affords the irreducible character $\zeta^{(2)}$ of Example 15.11.) Abbreviate S_4 as S, and write S as $A_4 \cup (12)A_4$. Induce the representation T by finding its values on the set of generators $\{(12), (13), (14)\}$ of S. Then, by taking appropriate products and using the conjugate classes as given in Example 15.12, find the character χ^S associated with T^S, and verify that χ^S is the irreducible character $\zeta^{(3)}$ of S_4.

19.2. The group $G = S_4$ has a dihedral subgroup P generated by the elements $x = (1423)$ and $y = (12)$. Now

$$T(x) = \begin{bmatrix} 0 & 1 \\ -1 & 0 \end{bmatrix}, \qquad T(y) = \begin{bmatrix} -1 & 0 \\ 0 & 1 \end{bmatrix}$$

determines a representation of P. Let χ be the character of T. Find χ^G, and reduce χ^G to its irreducible components.

19.3. The group $G = S_4$ has a normal subgroup H generated by $(12)(34)$ and $(13)(24)$ (see Example 15.12). Let χ be a character of degree 1 on H given by

$$\chi((12)(34)) = 1, \qquad \chi((13)(24)) = -1.$$

Let ψ be the character on H associated with the representation given by

$$U((12)(34)) = \begin{bmatrix} -1 & 0 \\ 0 & 1 \end{bmatrix}, \qquad U((13)(24)) = \begin{bmatrix} 1 & 0 \\ 0 & -1 \end{bmatrix}.$$

Find χ^G and ψ^G, and reduce each to its irreducible components.

19.4. Let $G = A_5$, $H = A_4$, and μ be the irreducible character $\zeta^{(4)}$ of A_4 (you could think of A_4 as the stabilizer of the point 5 in A_5). Use the Frobenius reciprocity theorem to reduce μ^G to its irreducible components.

19.5. Let $G = D_4 = \langle r, c : r^4 = c^2 = 1, cr = r^{-1}c \rangle$, let $H = \langle c \rangle$, and let T be the trivial representation of H given by $T(c) = [1]$ with character $\chi = \zeta^{(1)}$. Find the induced representation T^G and its character χ^G, and reduce χ^G to its irreducible components.

19.6. Prove the transitivity of induction, Theorem 19.10.

19.7. Prove that T^G, as defined in Definition 19.1, is a representation of G.

19.8.   Consider the trivial representation $T(h) = [1]$ for all elements h in
a subgroup H of some group G. The associated character is, of
course, the trivial character $\zeta^{(1)}$ with $\zeta^{(1)}(h) = 1$ for all $h \in H$.
Then T^G is a representation of G by permutation matrices, and the
degree of T is equal to the index $[G : H]$. Show that, if $h \in H$,
then $\chi^G(h) \geq [N_G H : H]$; then show also that if $H \trianglelefteq G$, then
$T^G(h) = I_{[G:H]}$ for all $h \in H$.

19.9.   For matrices $A = (a_{ij})$ and $B = (b_{ij})$ of degree n and m, respectively, we define the *tensor product* of A and B as

$$A \otimes B = \begin{bmatrix} a_{11}B & a_{12}B & \cdots & a_{1n}B \\ a_{21}B & a_{22}B & \cdots & a_{2n}B \\ & & \cdots & \\ a_{n1}B & a_{n2}B & \cdots & a_{nn}B \end{bmatrix},$$

where $a_{ij}B$ denotes the scalar multiple of the (sub)matrix B by a_{ij}.
If T and U are representations of a group G, we define the *tensor
product $T \otimes U$* by

$$(T \otimes U)(g) = T(g) \otimes U(g)$$

for each $g \in G$. For the representations T and U of S_4 given by

$$T((1234)) = \begin{bmatrix} 0 & -1 & 0 \\ 1 & 0 & 0 \\ 0 & 0 & 1 \end{bmatrix}, \quad T((243)) = \begin{bmatrix} 0 & -1 & 0 \\ 0 & 0 & -1 \\ 1 & 0 & 0 \end{bmatrix},$$

$$U((1234)) = \begin{bmatrix} 0 & \zeta^2 \\ \zeta & 0 \end{bmatrix}, \quad U((243)) = \begin{bmatrix} \zeta & 0 \\ 0 & \zeta^2 \end{bmatrix},$$

where ζ is a primitive cube root of 1, find $T \otimes U$ and its associated
character χ, and reduce χ to its irreducible components.

19.10. For the representation T of Exercise 19.9, find $T \otimes T$ and its associated character, and reduce that character to its irreducible components.

19.11. Let T and U be representations of a group G with associated characters χ and μ, respectively. Prove that the character of $T \otimes U$ as
defined in Exercise 19.9 is $\chi\mu$, specifically, show that the character
is

$$(\chi\mu)(g) = \chi(g)\mu(g)$$

for each $g \in G$.

Chapter 20

The Character Table for S_5

Having completed the character table for the group A_5 in Chapter 19, we are ready to extend that result one step farther, to S_5. To do so, we'll need only one additional result beyond what was available to us in Chapter 19. First, we'll tabulate the conjugate classes of S_5.

20.1 Proposition: The conjugate classes of S_5

The symmetric group S_5 of order 120 has the following conjugate classes:

C_1	$\{1\}$	$h_1 = 1$
C_2	elements of the form $(uv)(wx)$	$h_2 = 15$
C_3	elements of the form (uvw)	$h_3 = 20$
C_4	elements of the form $(uvwxy)$	$h_4 = 24$
C_5	elements of the form (uv)	$h_5 = 10$
C_6	elements of the form $(uvwx)$	$h_6 = 30$
C_7	elements of the form $(uv)(wxy)$	$h_7 = 20$

Proof: Classes C_1 through C_4 are from A_5; these classes remain distinct because no two contain elements of the same order. Since $A_5 \trianglelefteq S_5$, no other elements of S_5 can conjugate into these first four classes. Since C_5 through C_7 contain elements of order 2, 4, and 6, respectively, at least this many are required. The proof that all transpositions are in the class C_5 is the same as that given in Exercise 15.4, and the arguments for C_6 and C_7 are like unto it. $\square$

Since $A_5 \trianglelefteq S_5$, $S_5/A_5 \cong Z_2$, S_5 is nonabelian, and A_5 is simple, we know that $A_5 = S_5'$, and S_5 has exactly two characters of degree 1. It follows from Corollary 15.8 that S_5 must have at least one irreducible character of degree greater than 4, and from Theorem 19.8 and Example 19.11 that no irreducible character of S_5 can have degree greater than 10. You could try possible degrees until you come up with the solutions in integers to the equation

$$\sum_{i=3}^{7} z_i^2 = 118,$$

subject to the condition

$$2 \leq z_3 \leq z_4 \leq z_5 \leq z_6 \leq z_7 \leq 10;$$

however, there are four such solutions, namely, the sets $\{2, 2, 2, 5, 9\}$, $\{2, 2, 5, 6, 7\}$, $\{2, 3, 4, 5, 8\}$, and $\{4, 4, 5, 5, 6\}$, and you would still have to find out which of these was the correct one. Instead, we'll introduce (without proof) a more powerful technique that applies to S_n for any n.

20.2   Definition: Young diagrams

For given n, the associated *Young diagrams* (named for Alfred Young, who introduced them) consist of n squares arranged in k rows ($1 \leq k \leq n$) with each row beginning at the left of the diagram and with no row containing more squares than any row above it.

Examples of Young diagrams are as follows. For $n = 3$, we have

For $n = 4$, they are

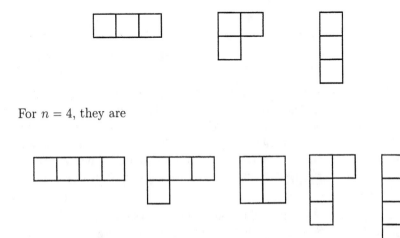

20.3 Definition: Hooks in Young diagrams

With each square in a Young diagram is associated a *hook* going from the center of the square to the right-hand edge of the diagram and also to the bottom of the diagram. The *length* of a hook is the number of squares through which the hook passes.

Sample hooks for two of the diagrams shown for $n = 4$ are as follows:

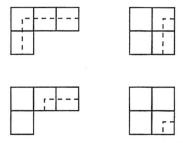

The lengths of the hooks of the Young diagrams are associated with the degrees of the irreducible characters of S_n by the following striking result.

20.4 Theorem: Degrees of the irreducible characters of S_n

Given n, for each Young diagram find the hook length associated with each square; then S_n has an irreducible character whose degree is $n!$ divided by the product of all of the hook lengths for the diagram.

In Examples 15.9 and 15.12 we found the character tables for S_3 and S_4, so we can use these groups as illustrations of the theorem. Here are the Young diagrams for $n = 3$ with the hook lengths shown in the respective squares:

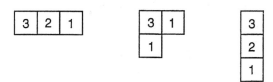

Now we have the degrees of the irreducible characters:

$$z_1 = \frac{3!}{3 \cdot 2 \cdot 1} = 1, \qquad z_2 = \frac{3!}{3 \cdot 2 \cdot 1} = 1, \qquad z_3 = \frac{3!}{3 \cdot 1 \cdot 1} = 2,$$

matching those in Example 15.9. Exercise 20.1 asks you to do the same for the irreducible characters of S_4. We'll proceed here to apply the theorem to S_5.

20.5 Proposition: Degrees of the irreducible characters of S_5

The irreducible characters for S_5 have the degrees 1, 1, 4, 4, 5, 5, and 6.

Proof: The Young diagrams for S_5 with the hook lengths for the squares are shown below; the degrees are then 5! divided by the products of the lengths for the diagrams.

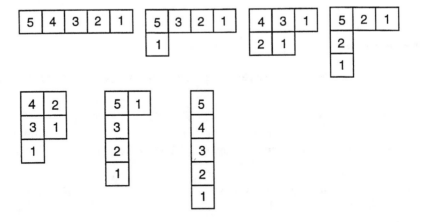

Noting, as we did in the proof of Proposition 20.1 that A_5 is the union of the first four conjugate classes listed for S_5, we immediately have the one-dimensional characters of S_5:

	C_1	C_2	C_3	C_4	C_5	C_6	C_7
$\zeta^{(1)}$	1	1	1	1	1	1	1
$\zeta^{(2)}$	1	1	1	1	-1	-1	-1

To complete the character table, we'll first use three induced characters, then appeal to the orthogonality relations. First, consider the irreducible character $\mu = \zeta^{(2)}$ of A_5 from Example 19.11. Write $G = S_5 = A_5 \cup (15)A_5$; by the methods of Chapter 19, you should obtain (see Exercise 20.2)

	C_1	C_2	C_3	C_4	C_5	C_6	C_7
μ^G	6	-2	0	1	0	0	0

Now $\deg(\mu) = 6$, so in view of Proposition 20.5, if μ reduces at all, it must have $\zeta^{(1)}$ or $\zeta^{(2)}$ as a constituent. But

$$\langle \mu^G, \zeta^{(1)} \rangle = \frac{1}{120}(6 - 30 + 24) = 0,$$

and similarly, $\langle \mu^G, \zeta^{(2)} \rangle = 0$; therefore, μ^G is the irreducible character of S_5 having degree 6, which we'll designate as $\zeta^{(7)}$.

Now let H be the stabilizer of the point 5 in the group S_5. Then $H \cong S_4$, and

$$G = S_5 = H \cup (15)H \cup (25)H \cup (35)H \cup (45)H,$$

where the coset $(15)H$ consists exactly of those permutations in S_5 that carry point 1 to point 5 (this is obvious from the fact that (15) carries 1 to 5, and any element of H then stabilizes 5), and similarly for the other cosets. Let

$$\phi = \zeta^{(1)} \qquad \text{and} \qquad \psi = \zeta^{(2)}$$

from the character table of S_4 (Example 15.12). Inducing the corresponding characters of G (Exercise 20.3), we have

	C_1	C_2	C_3	C_4	C_5	C_6	C_7
ϕ^G	5	1	2	0	3	1	0
ψ^G	5	1	2	0	-3	-1	0

but bear in mind that these characters need not be irreducible. In fact,

$$\langle \phi^G, \zeta^{(1)} \rangle = \frac{1}{120}(5 + 15 + 40 + 30 + 30) = 1,$$

$$\langle \phi^G, \zeta^{(2)} \rangle = \frac{1}{120}(5 + 15 + 40 - 30 - 30) = 0,$$

so ϕ^G is the sum of an irreducible character of degree 1 and (because of the degrees found in Proposition 20.5) a new irreducible character, of degree 4.

Similarly, ψ^G is the sum of the irreducible character $\zeta^{(2)}$ and yet another new irreducible character. Setting

$$\zeta^{(3)} = \phi^G - \zeta^{(1)} \qquad \text{and} \qquad \zeta^{(4)} = \psi^G - \zeta^{(2)},$$

we can add to our table for S_5

	C_1	C_2	C_3	C_4	C_5	C_6	C_7
$\zeta^{(3)}$	4	0	1	-1	2	0	-1
$\zeta^{(4)}$	4	0	1	-1	-2	0	1

From here on we can complete the table (that is, find the two irreducible characters of degree 5) by the orthogonality relations. For example, let

$$\zeta_2^{(5)} = a \qquad \text{and} \qquad \zeta_2^{(6)} = b;$$

then application of Corollary 15.7 to columns 2 and 1 gives $a + b = 2$, while applying the same corollary to the product of column 2 with itself gives $a^2 + b^2 = 2$. Solving this pair of equations gives $a = b = 1$. Exercise 20.4 asks you to complete this calculation, which completes the following.

20.6   Theorem: The character table for S_5

The character table for the symmetric group of order 120 is

	C_1	C_2	C_3	C_4	C_5	C_6	C_7
h_j	1	15	20	24	10	30	20
$\zeta^{(1)}$	1	1	1	1	1	1	1
$\zeta^{(2)}$	1	1	1	1	-1	-1	-1
$\zeta^{(3)}$	4	0	1	-1	2	0	-1
$\zeta^{(4)}$	4	0	1	-1	-2	0	1
$\zeta^{(5)}$	5	1	-1	0	1	-1	1
$\zeta^{(6)}$	5	1	-1	0	-1	1	-1
$\zeta^{(7)}$	6	-2	0	1	0	0	0

Exercises

20.1. Use Young diagrams to find the degrees of the irreducible characters of S_4, and verify that these agree with those in the character table of Example 15.12.

20.2. For the irreducible character $\mu = \zeta^{(2)}$ of A_5 from Example 19.11 and $G = S_5$, find the induced character μ^G (which is the $\zeta^{(7)}$ found in the text).

20.3. For the characters ϕ and ψ given in the text for S_4, find the induced characters on S_5 and verify that $\langle \psi^G, \zeta^{(1)} \rangle = 0$, $\langle \psi^G, \zeta^{(2)} \rangle = 1$.

20.4. Use Corollaries 15.7 and 15.5 to complete the character table for S_5 by finding $\zeta_j^{(i)}$ for $i = 5, 6$ and $j = 3$, 4, 5, 6, and 7.

20.5. Verify that Corollary 15.5 holds for the characters $\zeta^{(3)}$ and $\zeta^{(7)}$ found in the text for S_5.

20.6. Let χ be the character on S_5 that assigns to each element of the group the number of points in the set $\{1, 2, 3, 4, 5\}$ left fixed by that element. Find χ and reduce it to the sum of its irreducible components.

Chapter 21

Space Groups and Semidirect Products

In Application 3.15 we considered the "full" cubic group G^* of order 48, contrasted with the group G of the rigid cube in Application 3.14, which has order 24. Mathematical treatments usually regard the cube as solid and hence use the smaller group of symmetries; however, since a cubic group arising in crystallography allows the inversion center (or, equivalently, reflection in a plane parallel to two opposite faces of the cube), it will be helpful to see how the character table for the larger group can be formed from the table for the smaller one. Our first project in this chapter will be to prove a theorem that determines the connection between the two tables. The proof will be organized by means of the following three lemmas.

21.1 Lemma: Conjugate elements of subgroup and group

Let $H \leq G$, $[G : H] = 2$, a and b be elements of H, and $z \in C(G)$ with $z \notin H$. Then a and b are conjugate in H if and only if az and bz are conjugate in G.

Proof: If a and b are conjugate in H, then there exists $h \in H$ such that $h^{-1}ah = b$. Then $h^{-1}ahz = bz$, and since $z \in C(G)$, we have $h^{-1}azh = bz$. Conversely, if az and bz are conjugate in G, then there exists $g \in G$ such that $g^{-1}azg = bz$. Since $z \in C(G)$, we have $g^{-1}agz = bz$, and by cancellation, $g^{-1}ag = b$. Now since $[G : H] = 2$ and $z \notin H$, we have $G = H \cup Hz$, so either $g \in H$, in which event a and b are conjugate in H, or else $g = hz$ for some $h \in H$. In the latter event, $g^{-1}ag = (hz)^{-1}a(hz) = h^{-1}ah$ since $z \in C(G)$, whence a and b are conjugate in H. $\square$

21.2   Lemma: Conjugate classes of subgroup and group

Let $H \leq G$, $[G : H] = 2$, $z \in C(G)$, $z \notin H$, and $C_1, \ldots, C_s$ be the distinct conjugate classes of H. Then the distinct conjugate classes of G are

$$C_1, \ldots, C_s, C_1 z, \ldots, C_s z,$$

where $C_i z = \{hz : h \in C_i\}$ for $1 \leq i \leq s$.

Proof: We show first that the classes C_i of H remain distinct conjugate classes when considered as subsets of G. Let $a \in C_i$ and let $b \in C_j$, and suppose that there exists $h \in H$ such that

$$b = (hz)^{-1} a (hz).$$

Then $b = h^{-1} ah$ since $z \in C(G)$, and hence $i = j$; thus $C_1, \ldots, C_s$ are contained in s distinct conjugate classes of G. By Lemma 21.1 the subsets $C_1 z, \ldots, C_s z$ are also contained in s distinct conjugate classes of G, so the only question remaining is whether or not some C_i and some $C_j z$ might be in the same class. But $H \unlhd G$ by Proposition 4.18, and $C_i \subseteq H$, so no element of C_i can be conjugated out of C_i by an element of G; hence, $C_i \cap C_j z = \emptyset$ for every i and j. Thus the $2s$ subsets listed are all distinct conjugate classes, and since $G = H \cup Hz$, they account for all of the elements of G. $\square$

The next lemma and the theorem itself require an additional hypothesis (for the reason, see Exercise 21.7): that $z^2 = 1$.

21.3   Lemma: Irreducible characters from subgroup to group

Let $H \leq G$, $[G : H] = 2$, $z \in C(G)$, $z \notin H$, $z^2 = 1$; for $1 \leq i \leq s$, let $\zeta^{(i)}$ be an irreducible character of H and T_i be the representation of H affording $\zeta^{(i)}$. Define T_i^* and $T_i^\#$ on G by

$$T_i^*(hz^k) = T_i(h) \qquad \text{for } k = 0, 1;$$
$$T_i^\#(hz^k) = (-1)^k T_i(h) \qquad \text{for } k = 0, 1.$$

Then each T_i^* and each $T_i^{\#}$ are an irreducible representation of G. More-over, if χ_i^* and $\chi_i^{\#}$ are the respective characters, then

$$\chi_i^*(hz^k) = \chi(h) \qquad \text{for } k = 0, 1;$$
$$\chi_i^{\#}(hz^k) = (-1)^k \chi_i(h) \qquad \text{for } k = 0, 1.$$

Proof: Since for k and j each equal to 0 or 1, and for $h_1, h_2 \in H$,

$$\begin{aligned} T_i^*((h_1 z^k)(h_2 z^j)) &= T_i^*(h_1 h_2 z^{k+j}) \\ &= T_i(h_1 h_2) = T_i(h_1) T_i(h_2) \\ &= T_i^*(h_1 z^k) T_i^*(h_2 z^j), \end{aligned}$$

each T_i^* is a representation of G, and similarly for each $T_i^{\#}$. To verify that T_i^* is well defined, we need only to observe that if $h_1 z^k = h_2 z^j$, then $h_2^{-1} h_1 = z^{j-k}$, and since $z \notin H$, $h_1 = h_2$ and $k = j$. If T_i^* is reducible, then the nonsingular matrix A that reduces T_i^* will also reduce T_i; hence T_i^* is irreducible, and similarly for $T_i^{\#}$. The computation of the characters is obvious. $\qquad\square$

Now by Lemmas 21.2 and 21.3 we have proved the following.

21.4 Theorem: Character table for G^* from that for G

If $H \leq G$, $[G : H] = 2$, $z \in C(G)$, $z \notin H$, and $z^2 = 1$, then the character table for G is

C_1	$\cdots$	C_s	$C_1 z$	$\cdots$	$C_s z$
	$\zeta_j^{(i)}$			$\zeta_j^{(i)}$	
	$\zeta_j^{(i)}$			$-\zeta_j^{(i)}$	

where $(\zeta_j^{(i)})_{1 \leq i, j \leq s}$ is the character table for H.

To apply Theorem 21.4 to the two groups G^* and G of the cube (as in Applications 3.15 and 3.14, respectively), we need an element $z \in C(G)$ such that $z \notin G$. The element z of Application 3.15 satisfies the hypotheses of the theorem (see Exercise 21.8). Therefore, G^* has exactly 10 irreducible characters, and the table for G^* may be obtained from the character table of G determined in Example 15.12.

In view of the correspondence worked out in Example 16.6, we may remark that the effect of z above is to reflect each vertex of the cube through the center of the cube (as suggested in the text of Application 3.15); in applications z is called an *inversion*, with the center of the cube as an *inversion center*. Note that z is not a *rigid* motion in the sense of one that can be performed on a solid geometric cube without disassembling the object; in fact, as we said in Application 3.15, the effect of z upon a hollow cube is to turn the figure inside out. On the other hand, if one thinks of a cube as consisting only of its eight vertices (as in the symmetry of crystalline cesium chloride CsCl, in which each cesium ion is located at the center of a cube whose vertices are chloride ions), the motion z has physical meaning.

A similar analysis for the square bipyramid has been left as Exercise 21.1.

The idea of regarding a cube as consisting only of its vertices, as one does in interpreting the vertices as atoms in a crystal, leads us to the concepts of point and space groups used in chemistry. We imagine an infinite (or potentially infinite) lattice of points in the plane or in space, such as the points (x, y) in $\mathbf{R}^2$ or (x, y, z) in $\mathbf{R}^3$ having *integer* coordinates. These examples give square and cubical configurations, respectively, but, of course, others are possible.

A *space group* may be defined as that collection of geometric transformations that leaves a lattice of points invariant. These motions consist of rotations (about the origin), reflections, and translations. A point X of a lattice is moved to another point X' of the lattice by a matrix operation (a linear transformation specifying a rotation or a reflection) followed by a translation:

$$X' = RX + t. \tag{1}$$

The matrix operation R is called a *point operation*. We may represent transformation 1 in the Seitz notation $\{\hat{R} \mid \vec{t}\}$ used in much of the chemical literature, or more briefly (for our purposes) by (R, t). The operation in the space group is given by

$$(R, t)(S, u) = (RS, Ru + t); \tag{2}$$

here RS is again a point operation, and $Ru + t$ is another translation. Note that Ru itself is truly a translation; it moves the lattice by the length of u in the direction of the vector to which the rotation R carries the translation vector u. For example, if u is a unit vector in the direction of the positive X-axis and R is a rotation through 90° carrying the positive X-axis to the positive Y-axis, then Ru is a translation of one unit in the positive Y-direction.

The identity of a space group will be represented by $(E, 0)$, where E is the identity transformation and 0 is a translation through distance 0. To check the associative property, we calculate

$$
\begin{aligned}
[(Q, t)(R, u)](S, v) &= (QR, Qu + t)(S, v) \\
&= (QRS, QRv + Qu + t) \\
&= (Q, t)(RS, Rv + u) \\
&= (Q, t)[(R, u)(S, v)].
\end{aligned}
$$

Here, of course, we have used the group properties for the first and second coordinates. The inverse of (R, t) is $(R^{-1}, -R^{-1}t)$ since

$$
(R, t)(R^{-1}, -R^{-1}t) = (RR^{-1}, R(-R^{-1})t + t) = (E, 0)
$$

and

$$
\begin{aligned}
(R^{-1}, -R^{-1}t)(R, t) &= (R^{-1}R, R^{-1}t - R^{-1}t) \\
&= (E, 0).
\end{aligned}
$$

Thus we have (under the assumptions of closure) a space group G, which in fact is a group acting on the set of lattice points in the sense of Definition 3.1.

We choose a set of *basic primitive translations*, one for each dimension of the $\mathbf{R}^n$ under consideration. A linear combination of basic primitive translations using integer coefficients is called a *primitive translation*. Because the point operations permute the points of the lattice, Ru will be a primitive translation whenever R is a point operation and u is a primitive translation.

We'll simplify our discussion by restricting ourselves to lattices in the plane $\mathbf{R}^2$. We begin with a lattice L consisting of all points (x, y) with integer coordinates. The basic primitive translations t_1 and t_2 will be movements of the entire lattice L one unit to the right and one unit up, respectively. Thus, for example, $t_1 + t_2$ will move the lattice $\sqrt{2}$ units up the line $x = y$. For point operations we consider a rotation C_4 through $90°$ counterclockwise around the (fixed) origin. (Here C_n is the usual symbol in chemical applications for a rotation through $2\pi/n$; thus $C_4^2 = C_2$ and $C_4^4 = E$.)

The operation (C_4, t_1) rotates the lattice through $90°$ and then shifts it one unit to the right. The elements (C_4, t_1) and $(C_4, t_1)^2$ may be seen in Figure 21.1 for nine points of the lattice L. Note that by Equation 2 above, we have

$$
(C_4, t_1)(C_4, t_1) = (C_4^2, C_4 t_1 + t_1),
$$

where $C_4 t_1 = t_2$, as described following Equation 2. You should verify geometrically that $(C_4, t_1)^4 = (E, 0)$. Further examples are given in Exercise 21.3.

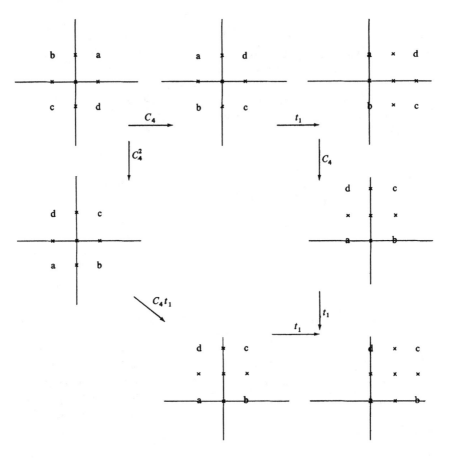

FIGURE 21.1
Rotation and translation in plane

Now the set of primitive translations forms a subgroup of G, as is clear from the computation

$$(E,t)(E,u) = (E, Eu + t) = (E, u + t)$$

for closure and from the existence of an inverse $(E, -t)$ for (E, t) (see Proposition 2.8). Moreover, if R is a point operation and if t and u are primitive translations, then

$$\begin{aligned}
(R,t)^{-1}(E,u)(R,t) &= (R^{-1}, -R^{-1}t)(E,u)(R,t) \\
&= (R^{-1}, R^{-1}u - R^{-1}t)(R,t) \\
&= (E, R^{-1}t + R^{-1}u - R^{-1}t) \\
&= (E, R^{-1}u);
\end{aligned}$$

hence, the set T of all primitive translations forms a normal subgroup of G.

The point operations of a space group will form a group P, called the *point group*, which is isomorphic to G/T, but which in general *need not be a subgroup of G*. Even if P is a subgroup of G, so that we can write typical elements in the form $(R, 0)$, we still have

$$
\begin{aligned}
(S, t)^{-1}(R, 0)(S, t) &= (S^{-1}, -S^{-1}t)(R, 0), (S, t) \\
&= (S^{-1}R, -S^{-1}t)(S, t) \\
&= (S^{-1}RS, S^{-1}Rt - S^{-1}t),
\end{aligned}
$$

which will not in general have $S^{-1}Rt - S^{-1}t = 0$; thus P is not normal in G, and in particular, G is not the direct product of P and T. A space group in which the point group is a (not necessarily normal) subgroup is called *symmorphic*; we'll examine a mathematical characterization of symmorphic space groups later in this chapter.

Next let's find the irreducible representations of a group T of primitive translations. To avoid the problems of infinite space groups and to avoid surface problems (due to the fact that lattice sites at a surface of a crystal are obviously in a different environment from those in the bulk), we'll resort to the use of the *cyclic boundary condition*: once one reaches a boundary of a crystal (or plane figure), one immediately moves to a corresponding location on the opposite boundary. For each of the basic primitive translations t_1, t_2 in $\mathbf{R}^2$ or t_1, t_2, t_3 in $\mathbf{R}^3$, we assume that there is an integer N_j such that $(E, t_j)^{N_j} = (E, 0)$ with $j = 1, 2, 3$. Equivalently, $(E, N_j t_j) = (E, 0)$ for each j. Then the dimensions of the figure or crystal are

$$
N_1|t_1| - \text{by} - N_1|t_2|(-\text{by} - N_3|t_3|).
$$

The group T of primitive translations is obviously abelian, and thus is the direct product of the $\langle (E, t_i) \rangle$, where the t_i are the basic primitive translations. By Theorems 13.12 and 14.15, the irreducible representations are all one dimensional, and there are $N_1 N_2$ or $N_1 N_2 N_3$ of them (accordingly, as the dimension of the space is 2 or 3). Based on the relations given by the assumption of cyclic boundary conditions, the irreducible representations of T are those found in the proof of Theorem 13.12 with $n = 2$ or 3, ζ_j a primitive N_jth root of 1, and $0 \le b_j < N_j$. Specifically, *one* such irreducible representation is

$$
U\left(\sum_{j=1}^n a_j t_j\right) = \prod_{j=1}^n (\zeta_j^{b_j})^{a_j} \qquad \text{for } 0 \le a_j < N_j.
$$

The general form of a representation U of T is sometimes given in the chemical literature (for each $a_j = 1$) as

$$\exp\left[i\frac{2\pi b_j}{t_j N_j}\cdot t_j\right] \qquad \text{for } 0 \le b_j \le N_j - 1,$$

where $\exp[A]$ denotes e^A and the dot denotes the scalar (dot) product of vectors.

Now the reducible representations of T may be formed from the U above in the usual way, by taking linear combinations with integer coefficients of the irreducible representations.

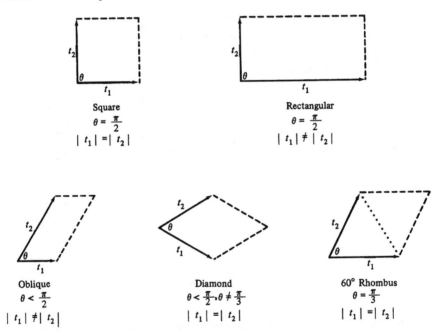

FIGURE 21.2
Plane lattice units

We'll turn next to some specific space groups for lattices in $\mathbf{R}^2$. The five possible lattice units are shown in Figure 21.2. We'll consider two of the seven groups associated with a rectangular lattice; the others are treated in standard references such as the [*International Tables for X-Ray Crystallography*, Vol. 1]. In Figure 21.3 are shown the configurations that must be preserved by point operations and translations; each ⊢ or ⌊ must be carried into another ⊢ or ⌊. The symbols *pm* and *pg* are often used to denote the groups of these two configurations (here *p* stands for "primitive," m for "mirror plane," and g for "glide plane"). Each of the groups pm and pg has a point group that is cyclic of order 2.

The group pm has a reflection σ_v generating its point group; it also has the basic primitive translations t_1 and t_2. Thus we may regard pm as generated by the elements $(\sigma_v, 0)$, (E, t_1), and (E, t_2). By inspection, you can see that $(\sigma_v, 0)$ commutes with (E, t_1) and that

$$(\sigma_v, 0)(E, t_2) = (E, -t_2)(\sigma_v, 0).$$

In order to write pm as an *abstract* group as in Chapter 5, we must change the primitive translation group T to multiplicative notation, writing $t_1 t_2$ instead of $t_1 + t_2$, and t_1^{-1} in place of $-t_1$. Then we may summarize our discussion of the structure of pm as

$$pm = \langle s, t, u : s^2 = 1, st = ts, su = u^{-1}s, tu = ut \rangle,$$

where we have written s for σ_v, t for t_1, and u for t_2. Of course, this characterization makes pm infinite; if we adopt cyclic boundary conditions on t and u, we have

$$pm = \langle s, t, u : s^2 = t^n = u^m = 1, st = ts, su = u^{-1}s, tu = ut \rangle,$$

for some (presumably large) integers n and m. Either way, $T = \langle t, u \rangle$ forms a normal subgroup, as shown before, since conjugation of the generators t and u of T gives an element of T, by the defining relations. You should note, however, that $\langle s \rangle$ is not normal since

$$u^{-1}su = su^2 \notin \langle s \rangle.$$

The group pg has translations t_1 and t_2, and also a *glide plane* $(\sigma_v, \frac{1}{2}t_1)$, as shown in Figure 21.3. Now by inspection of the geometry of the figure, we see that

$$\left(\sigma_v, \frac{1}{2}t_1 \right)^2 = (E, t_1)$$

and that

$$\left(\sigma_v, \frac{1}{2}t_1 \right)(E, t_2) = (E, -t_2)\left(\sigma_v, \frac{1}{2}t_1 \right),$$

which may be described in words as

> The product of two glide plane operations is a basic primitive translation in the horizontal direction,

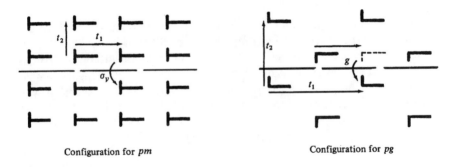

Configuration for *pm*   Configuration for *pg*

FIGURE 21.3
Configurations for *pm* and *pg*

and

> A glide plane operation followed by a basic primitive translation *up*
> is the same as a primitive translation *down* followed by a glide plane
> operation.

Verifying the two relations given by means of the product rule Equation
2) leads to terms such as $\sigma_v \frac{1}{2} t_1$. Such a term can be viewed geometrically,
although it may seem less than intuitive in a chemical context because
the glide plane is composed of two inseparable parts: the reflection and the
half-translation (where the latter is not primitive). The problem is resolved
when we change the primitive translation subgroup into multiplicative notation and think of *pg* as an abstract group. Let g denote the glide plane
operation and u the basic primitive translation upward; then our previous
calculations give the relation $ug = gu^{-1}$ and the fact that g^2 is the basic
primitive translation to the right, so we need only these two generators.
Thus we may write

$$pg = \langle g, u : ug = gu^{-1} \rangle,$$

or, assuming cyclic boundary conditions, we take (large) integers n and m
with

$$pg = \langle g, u : g^{2n} = u^m = 1, ug = gu^{-1} \rangle.$$

It is important to realize here that the subgroup T of primitive translations
is generated by g^2 and u. Now the point group pg/T is cyclic of order 2,
and $pg = T \cup Tg$, but g does *not* generate a subgroup of order 2 in *pg*.

Thus we conclude that *pm* is symmorphic and *pg* is nonsymmorphic.
Once we have introduced the idea of semidirect products, we'll be able to
say even a bit more about the group *pg*.

As a prelude to the final topic of this chapter, recall that at the end of
Chapter 6 we mentioned the study of group extensions, in which, given the

kernel N and an isomorphic copy H of the factor group G/N, we try to find one or more groups G that have the given kernel and the given factor group. This is referred to as the *extension problem* for groups. (Strictly speaking, we ask that G contain an isomorphic copy of the given group N, but such niceties of terminology need not intrude upon the discussion here.) The extension problem has, as you know by now, a trivial answer for every N and H, namely, the external direct product $G = N \times H$ of Proposition 13.4. Now we'll find (in some cases) a more general product, one in which the subgroup H of the product of H and N is not necessarily a normal subgroup of the product (as it is in the direct product). To make the idea precise, we'll formulate the following definition.

21.5 Definition: Semidirect product

If $A \trianglelefteq G$, $H \leq G$, $G = HA$, and $H \cap A = \{1\}$, then G is called the *semidirect product* of H by A.

If either H or A is the subgroup $\{1\}$, then the decomposition of G as a semidirect product is trivial; hence we'll assume in general that H and A are both different from $\{1\}$. Incidentally, the literature of group theory is not entirely consistent about the use of the phrase "H by A" — some authors use a variant of this phrase, and you'll need to check any other text you read to see what convention is being used.

Directly from Theorem 6.11 we have the following.

21.6 Proposition: Factor group of a semidirect product

If G is the semidirect product of subgroups H by A (with $A \trianglelefteq G$), then $H \cong G/A$.

Let's illustrate the concept with a pair of examples. First, take

$$G = D_4 = \langle r, c : r^4 = c^2 = 1, cr = r^{-1}c \rangle,$$

and let $A = \langle r \rangle$ and $H = \langle c \rangle$. Then clearly G is the semidirect product of H by A. On the other hand, try taking $A = \langle r^2 \rangle$, the only *normal* subgroup of order 2. Then there is *no* subgroup H for which we can write G as the semidirect product of H by A since A is contained in every subgroup of order 4 (see Figure 6.1 in Chapter 6).

For another example, consider

$$Q_2 = \langle x, y : x^4 = 1, y^2 = x^2, yx = x^{-1}y \rangle.$$

Now Q_2 has a unique subgroup $\langle x^2 \rangle$ of order 2, and it is contained in each of the subgroups $\langle x \rangle$, $\langle y \rangle$, and $\langle xy \rangle$ of order 4; hence, Q_2 cannot be written as the semidirect product of a subgroup of order 4 by a normal subgroup of order 2. In addition, since Q_2 has only the subgroups listed above (together with itself and $\{1\}$), it cannot be written as the semidirect product of a subgroup of order 2 by a normal subgroup of order 4. In other words, the *only* way in which to write Q_2 as a semidirect product is to take either H or A to be $\{1\}$, the trivial case, which we are ignoring.

An immediate consequence of Definition 21.5 is the observation that every direct product is also a semidirect product. The example of D_4 given above shows that a semidirect product need not be a direct product.

Recall from Proposition 5.6 that the automorphisms of a group form a group under the operation of composition. If G is the semidirect product of H by A, we'll need to consider the group $\mathbf{A}(A)$ of automorphisms of A in the following basic result on semidirect products.

21.7   Proposition: Properties of a semidirect product

Let G be the semidirect product of H by A (with $A \trianglelefteq G$). Then,

(a)   Each $g \in G$ can be written uniquely as a product ha with $h \in H$ and $a \in A$;

(b)   If $h_1, h_2 \in H$ and $a_1, a_2 \in A$, then

$$(h_1 a_1)(h_2 a_2) = (h_1 h_2)[(h_2^{-1} a_1 h_2) a_2];$$

(c)   The function $\phi : H \rightarrow \mathbf{A}(A)$ given by $\phi(h) : a \rightarrow hah^{-1}$ is a homomorphism;

(d) $\phi(h)$ in part c need not be an inner automorphism of A.

Proof of a: Since $G = HA$, we know that each element of G can be written as ha, with $h \in H$ and $a \in A$. If $ha = h'a'$, then

$$(h')^{-1}h = a'a^{-1} \in H \cap A,$$

so $(h')^{-1}h = 1$, whence $h' = h$; similarly, $a' = a$.

Proof of b: Since $A \trianglelefteq G$, we have

$$(h_1 a_1)(h_2 a_2) = h_1 (h_2 h_2^{-1}) a_1 h_2 a_2$$
$$= (h_1 h_2)[(h_2^{-1} a_1 h_2) a_2] \in HA.$$

Proof of c: If $h, h' \in H$, then

$$\phi(hh') : a \to (hh')a(hh')^{-1} = h(h'ah'^{-1})h^{-1}$$

and

$$\phi(h)\phi(h') : a \to h(h'ah'^{-1})h^{-1},$$

as required.

Proof of d: In the example above with $G = D_4$ and $A = \langle r \rangle$, A is abelian, so its only inner automorphism is the identity, but

$$\phi(c) : r \to crc^{-1} = crc = r^{-1},$$

which is not the identity automorphism. □

Recall from Proposition 13.4 that if G is the *direct* product of H by A, then every element in G can be written uniquely as a product ha with $h \in H$ and $a \in A$, and that the operation in G may be expressed as

$$(h_1 a_1)(h_2 a_2) = (h_1 h_2)(a_1 a_2).$$

To reconcile this product with the one in Proposition 21.7b, we need only to note that from Proposition 13.3ii, h_2 commutes with a_1 when the product is direct.

Given groups H and A, we can form an "external" semidirect product by analogy with Proposition 13.4. By changing notation, if necessary, we'll take $H \cap A = \{1\}$. First we let ϕ be a homomorphism from H into $\mathbf{A}(A)$ (noting that *each* $\phi(h)$ is then an automorphism of A), and set

$$a^h = \phi(h^{-1})a \qquad \text{for } a \in A, h \in H.$$

Then
$$(a^h)^k = (\phi(h^{-1})a)^k = \phi(k^{-1})\phi(h^{-1})a$$
$$= \phi((hk)^{-1})a = a^{hk} \qquad \text{for } h, k \in H;$$
$$a^1 = \phi(1)a = a;$$

hence by Definition 3.1 it follows that we have defined an action of H on A. Now we take G to be the set of formal products of the form ha with $h \in H$ and $a \in A$. (No other operation than juxtaposition is assumed between h and a.) For $h, k \in H$ and $a, b \in A$, we define the product by

$$(ha)(kb) = (hk)(a^k b) \in HA = G.$$

Clearly, $(ha)(1 \cdot 1) = h(a^1) = ha = (1 \cdot 1)(ha)$, so G has the identity element $1 \cdot 1$, which we'll write simply as 1. Since $\phi(h^{-1})$ is an automorphism of A, there is a unique solution b to the equation $a^{h^{-1}} b = 1$ in A, and for this b, $h^{-1}b$ is an inverse for ha. For associativity, observe that if $h, k, m \in H$ and $a, b, c \in A$, then

$$[(ha)(kb)](mc) = [(hk)(a^k b)](mc) = (hkm)(a^k b)^m c$$
$$= (hkm)(a^{km} b^m c) = (ha)[(km)(b^m c)]$$
$$= (ha)[(kb)(mc)],$$

as required. Hence, G is a group with the specified product. If we identify A with the set of elements $1a$ (abbreviated a) and H with the set of elements $h1$ (abbreviated h), then A and H are clearly subgroups of G. Finally, for $h \in H, a \in A,$

$$hah^{-1} = (h^{-1}1)(1a)(h1) = [h^{-1}(1^1 a)](h1)$$
$$= (h^{-1}a)(h1) = 1a^h \in A,$$

so $A \trianglelefteq G$. This completes the proof that G is the semidirect product of H by A. This external product may be related to the internal one by comparing the ϕ here with the ϕ of Theorem 21.7c.

Note that if in the equation $a^h = \phi(h^{-1})a$, $\phi(h^{-1})$ is the identity function for every $h \in H$, then the result in the construction of the external semidirect product is simply the direct product. In some instances, the *only* homomorphism from H to $\mathbf{A}(A)$ is the trivial one carrying every $h \in H$ to the identity function; in this event the only semidirect product of H by A is the direct product. For example, let $H = A \cong \mathbf{Z}_3$; then since $\mathbf{A}(\mathbf{Z}_3) \cong \mathbf{Z}_2$ and since when ϕ is a homomorphism, the order of an element $\phi(x)$ must divide the order of x, the only possible choice of ϕ in the construction is the trivial one. Then the only semidirect product of H by A must be the direct product $\mathbf{Z}_3 \times \mathbf{Z}_3$.

To conclude this discussion, observe first that pm, in either its finite form or its infinite form, is the semidirect product of $\langle s \rangle$ by $T = \langle t, u \rangle$. You can easily reconcile the different notations for the group product. Second, we'll show the following.

21.8 Proposition: *pg* is nonsymmorphic

The space group pg cannot be written as a semidirect product of some subgroup by $T = \langle g^2, u \rangle$.

Proof: We'll take $pg = \langle g, u : ug = gu^{-1} \rangle$ since even with the cyclic boundary condition we would assume that the orders of g and u were very large (for the lattice of atoms in a crystal). We already know that pg is not the semidirect product of $\langle g \rangle$ by T since $g^2 \in T$; the point is to show that there is *no* subgroup H of pg such that $pg = HT$, $H \cap T = \{1\}$. But $[pg : T] = 2$, so such an H would have to have order 2, and we know that no element of pg other than the identity has finite (or at least small finite) order. Therefore, pg cannot be written as a semidirect product with T as the normal subgroup. $\qquad\square$

Although the preceding result answers the question for chemical applications, as a point of interest we can actually prove a mathematically different statement.

21.9 Proposition: *pg* can be written as a semidirect product

The space group pg can be written as a semidirect product with $\langle u \rangle$ as the normal subgroup.

Proof: First, we'll show that $\langle u \rangle$ is indeed a normal subgroup of pg. Now from the relation $ug = gu^{-1}$,

$$g^{-1}ug = g^{-1}(gu^{-1}) = u^{-1} \in \langle u \rangle$$

and (as you can always do from an equation like the preceding)

$$g^{-1}u^{-1}g = (g^{-1}ug)^{-1} = (u^{-1})^{-1} = u \in \langle u \rangle;$$

hence $\langle u \rangle \trianglelefteq pg$. Now since $gu \neq ug$, pg cannot be the *direct* product of $\langle g \rangle$ and $\langle u \rangle$, but if we can show that $pg = \langle g \rangle \langle u \rangle$, it will follow (with $\langle g \rangle \cap \langle u \rangle = \{1\}$) that pg is the semidirect product of $\langle g \rangle$ by $\langle u \rangle$. A caution is in order here: the fact that $pg = \langle g, u \rangle$ does *not* imply immediately that $pg = \langle g \rangle \langle u \rangle$; we must check this fact. But since we have the defining relation $ug = gu^{-1}$, we see that any u in a product of powers of g and u can be moved to the right across any g^k with $k > 0$. From our calculation $g^{-1}u^{-1}g = u$, we have $ug^{-1} = g^{-1}u^{-1}$; and from $g^{-1}ug = u^{-1}$, we have $u^{-1}g^{-1} = g^{-1}u$. Hence, any u^{-1} in such a product may also be moved to the right across any g^k with $k < 0$. Hence, in any arbitrary product of powers of g and u, all powers of u may be collected on the right, leaving all powers of g on the left; that is, the product may be brought into the form $g^j u^k$ with integer exponents. Therefore, $pg = \langle g \rangle \langle u \rangle$. This completes the proof. $\square$

Bear in mind that *a space group is symmorphic if and only if it can be written as the semidirect product of its point group and its subgroup of primitive translations.* Thus pg remains nonsymmorphic despite Proposition 21.9.

_____

Exercises

21.1. The square bipyramid also has a full symmetry group G^* consisting of the group G found in Chapter 5 (note that $G \cong D_4$) and a coset Gz whose representative z is an inversion of all vertices through the center of the solid. Find the character table of G^*.

21.2. From the character tables found in Chapter 15, observe that Theorem 21.4 cannot hold with $H = A_4$, $G = S_4$. Determine why Theorem 21.4 does not apply, and discuss this result geometrically with respect to the fact that A_4 is the group of rigid symmetries of the tetrahedron.

21.3. In the notation of this section, verify geometrically that

$$C_4 t_2 = -t_1,$$

$$C_4 t_1 + t_2 = 2t_2,$$

$$(C_4, t_2)(C_4, t_1) = (C_4^2, 2t_2),$$

$$(C_4, t_1 - t_2)^{-1} = (C_4^3, -C_4^3(t_1 - t_2)).$$

21.4. We found in Proposition 21.9 that pg can be written as a semidirect product of $\langle g \rangle$ by $\langle u \rangle$. Determine whether or not, for *its* vertical basic primitive translation u, the group pm can be written as a semidirect product of some subgroup by $\langle u \rangle$. Can it be written as a *direct* product of some subgroup and $\langle u \rangle$?

21.5. We know that $S_4 = \langle r, h \rangle$ (as in the text following Theorem 21.4) but that $S_4 \neq \langle r \rangle \langle h \rangle$. Examine the proof of Proposition 21.9 and determine why you cannot carry out for S_4 the argument given there to show that $pg = \langle g \rangle \langle u \rangle$.

21.6. Determine the two distinct semidirect products of $\mathbf{Z}_2$ by $\mathbf{Z}_\infty$.

21.7. Let $G = \mathbf{Z}_8 = \langle g \rangle$, $H = \langle g^2 \rangle$, and $z = g$ in the notation of Lemma 21.3. Here $z^2 \neq 1$. Find an example of $\zeta^{(i)}$ for which the conclusion of Lemma 21.3 fails.

21.8. Verify that $z = (17)(28)(35)(46)$ in Application 3.15 satisfies the hypotheses of Theorem 21.4.

Chapter 22

Proofs of the Sylow Theorems

To complete the presentation of the material in Chapter 7, here are proofs of the Sylow theorems. The presentation here emphasizes the concept of a group acting on a set and is derived from that of Helmut Wielandt ("Zum Satz von Sylow," *Mathematische Zeitschrift* **60** (1954): 407–408).

We'll need two preliminary results; the first is easy to see, and the second should be familiar to you.

22.1 Remark: An action on the power set

Let Ω be the power set of G (the set of all subsets of G); and for $g \in G$ and $U \in \Omega$, let

$$U^g = \{ug : u \in U\}.$$

Then this operation is an action of G on Ω. Furthermore, since $|U^g| = |U|$ for any subset U of G, if Ω is, instead, the collection of all subsets of G having a given order k, then the given operation is again an action of G on this Ω.

22.2 Remark: Combinations

The number of distinct subsets of order m that can be made from a population of order n (usually referred to as "the combinations of n things taken m at a time") is given by the binomial coefficient:

$$\binom{n}{m} = \frac{n!}{m!(n-m)!}.$$

Now we are ready to prove the following.

22.3   Theorem: Sylow theorem I

Let p be a prime and $|G| = p^e q$, where p does not divide q. Then G contains a subgroup of order p^e, called a *Sylow p-subgroup*.

Proof: We'll begin by taking Ω in Remark 22.1 to be the collection of subsets of G having order p^e; we do not assume that any such subset is a subgroup! By Remark 22.2, the number of such subsets is

$$\binom{|G|}{p^e} = \frac{|G|(|G|-1)(|G|-2)\cdots(|G|-k)\cdots(|G|-(p^e-1))}{p^e \cdot 1 \cdot 2 \cdot \ldots \cdot k \cdot \ldots \cdot (p^e-1)}.$$

Now we'll show that whenever a power of p appears in the numerator, it cancels one in the denominator. First, $|G|/p^e$ at the beginning of the right-hand side reduces to q, which is not divisible by p. Now suppose that p^a divides $(|G|-k)$ in the numerator for some $a \geq 1$. If $a \geq e$, then p^e also divides $(|G|-k)$, and since p^e divides $|G|$, it also divides k, which contradicts the fact that $k \leq p^e - 1$. Hence $a < e$, which implies that p^a divides $|G|$ and consequently that p^a divides k, as was to be shown. Now on the one hand, any power of p in the numerator cancels one in the denominator, and on the other hand, it is well known that

$$\binom{n}{m}$$

is a positive integer; hence, p does not divide

$$\binom{|G|}{p^e},$$

which is $|\Omega|$.

We let G act on Ω by right multiplication, as in Remark 22.1, and let $\Omega_1, \ldots, \Omega_n$ be the distinct orbits of Ω under this action. For each $i = 1, \ldots, n$, let $\Omega_i = U_i^G$; that is, U_i is a typical element of the orbit Ω_i. Now

$$|\Omega| = \sum_{i=1}^{n} |\Omega_i|,$$

and if p divides *every* $|\Omega_i|$, then p divides $|\Omega|$, which is a contradiction. Hence there exists at least one i such that p does not divide $|\Omega_i|$. Suppose that $i = 1$. Let S be the stabilizer of the subset U_1; we'll show that S is a Sylow p-subgroup of G. We know that the stabilizer is a subgroup, so we need only to prove that it has order p^e.

If $g \in S$, then $U_1^g = U_1$, and if $u \in U_1$, then $ug \in U_1$. Hence

$$g \in u^{-1}U_1 = \{u^{-1}x : x \in U_1\}.$$

Therefore, $S \subseteq u^{-1}U_1$ and $|S| \le |u^{-1}U_1| = |U_1| = p^e$. On the other hand, we took $|\Omega_1|$ as not divisible by p, and by Theorem 3.13, $|\Omega_1| = |G|/|S|$. But Theorem 3.13 also says that $|\Omega_1|$ divides $|G| = p^e q$. Hence $|\Omega_1|$ must divide q and so is $\le q$; consequently $|S| = |G|/|\Omega_1| \ge p^e$. From the two inequalities, we conclude that $|S| = p^e$; this completes the proof of Theorem 22.3.   □

22.4   Theorem: Sylow theorem II

Let $|G| = p^e q$ with p prime and p not dividing q. Let $H \le G$ with $|H| = p^k$, and let P be *any* Sylow p-subgroup of G. Then there exists an element $g \in G$ such that $H \le g^{-1}Pg$.

Proof: Let S be the Sylow p-subgroup obtained in the proof of Theorem 22.3; S is the stabilizer of a subset U_1 of order p^e under the action of right multiplication on the subsets of G. We'll establish the result first under the assumption that $P = S$ and then generalize to the case of $P \ne S$.

Case 1. $P = S$. By restriction of the action in Remark 22.1 to the subgroup H, it follows that H acts on the set $\Omega_1 = U_1^G$. However, we shall expect Ω_1 to have more than one orbit under H since only those elements of G that are in H are available to carry subsets in Ω_1 to one another. Let $X_1, \ldots, X_m$ be the distinct orbits of Ω_1 under H, with $X_i = Y_i^H$ for each $i = 1, \ldots, m$. Note that each $Y_i \in \Omega_1$. Then by Theorem 3.13, $|X_i| = |H|/|H_{y_i}|$, and since $|H|$ is a power of p, there exist nonnegative integers $c_1, \ldots, c_m$ such that $|X_i| = p^{c_i}$ for each i. But

$$|\Omega_1| = \sum_{i=1}^{m} |X_i|$$

and p does not divide $|\Omega_1|$; hence, there exists at least one i such that p does not divide $|X_i|$, and we may assume that p does not divide $|X_1|$. Now $|X_1|$

is a power of p and is not divisible by p; this can occur only if $|X_1| = p^0 = 1$. Hence H is contained in the stabilizer of Y_1. Since $Y_1 \in \Omega_1$, there exists $g \in G$ such that $Y_1 = U_1^g$. Since S is the stabilizer of U_1, we conclude by Proposition 5.10 that H is contained in $g^{-1}Sg$. This completes the proof for the case in which $P = S$.

Case 2. $P \neq S$. By Case 1 (with $k = e$) we know that there exists $a \in G$ such that $P \leq a^{-1}Sa$. Now $|P| = |a^{-1}Sa|$, so $P = a^{-1}Sa$. Also by the first case, there exists $b \in G$ such that $H \leq b^{-1}Sb$. Let $g = a^{-1}b$; then

$$g^{-1}Pg = (b^{-1}a)(a^{-1}Sa)(a^{-1}b) = b^{-1}Sb,$$

and hence $H \leq g^{-1}Pg$. This completes the proof of Theorem 22.4.   □

22.5   Theorem: Sylow theorem III

Let $|G| = p^e q$ with p prime and p not dividing q. Then the number s_p of distinct Sylow p-subgroups of G is congruent to 1 modulo p and is a divisor of q.

Proof: First, Theorem 22.4 says that any two Sylow p-subgroups are conjugate and hence, for the given prime p, the set of Sylow p-subgroups of G has only one orbit under the action of conjugation: thus s_p is the order of that one orbit. If P is a Sylow p-subgroup of G and N is the normalizer of P, then $s_p = [G : N]$ by Theorems 3.12 and 3.13. Now $P \leq N$, so $|P|$ divides $|N|$ and hence by Theorem 3.12,

$$|G| = |N|[G : N] = |P|[G : P] \quad \text{and}$$
$$s_p = [G : N] \quad \text{divides} \quad [G : P] = q.$$

This completes the first part of the proof (the second part of the statement).

For the second part, we must show that there exists a nonnegative integer k such that $s_p = 1 + kp$. Let $\Omega = \{P_1, \ldots, P_{s_p}\}$ be the set of all Sylow p-subgroups of G (for our fixed p), and let P_1 act on the set Ω by conjugation. For each $i = 1, \ldots, s_p$, let N_i be the normalizer in P_1 of P_i; that is, N_i consists of those elements g of P_1 for which $g^{-1}P_ig = P_i$. Now Ω is the union of its distinct orbits $\Omega_1, \ldots, \Omega_r$ under P_1; let $k_j = |\Omega_j|$ for $j = 1, \ldots, r$. If $P_i \in \Omega_j$, then $k_j = [P_1 : N_i]$. By the closure property in a group, $N_1 = P_1$ and we may take $\Omega_1 = \{P_1\}$ with $k_1 = 1$. We'll show now that $k_j > 1$ for $2 \leq j \leq r$. If $k_j = 1$, then $\Omega_j = \{P_i\}$ for some i, and $P_1 = N_i$. Hence, $P_1P_i = P_iP_1$ and by Proposition 6.2, $P_1P_i \leq G$. Moreover, since P_1 and P_i

are both contained in $N_G P_i$, we have $P_i \trianglelefteq P_1 P_i$. By Theorem 6.11,

$$P_1 P_i / P_i \cong P_1/(P_1 \cap P_i);$$

hence by Theorem 3.12, $|P_1 P_i|$ is the product of $|P_i|$ and $|P_1/(P_1 \cap P_i)|$, which makes $|P_1 P_i|$ a power of p. But since P_i is a Sylow p-subgroup of G, $|P_1 P_i| = |P_i|$; hence, $P_1 = P_i$ and $i = 1$, whence $j = 1$. This establishes the fact that $k_j > 1$ for $2 \leq j \leq r$. Now $k_j = [P_1 : N_i]$ is a divisor of $|P_1|$ and hence a power of p. Thus p divides the sum $k_2 + \cdots + k_r$, and we conclude that there is an integer $k \geq 0$ such that

$$\begin{aligned} s_p &= k_1 + (k_2 + \cdots + k_r) \\ &= 1 + kp. \end{aligned}$$

This completes the proof of all three Sylow theorems. □

Bibliography

Adams, Jeffrey, *Representation Theory of Groups and Algebras*, Providence, RI: American Mathematical Society, 1993.

Alperin, J. L. and Rowen B. Bell, *Groups and Representations*, New York: Springer-Verlag, 1995.

Altmann, Simon L., *Band Theory of Solids: An Introduction from the Point of View of Symmetry*, Oxford: Oxford University Press, 1990.

Aschbacher, Michael, *Finite Group Theory*, Cambridge: Cambridge University Press, 1988.

Brumfiel, Gregory W. and H. M. Hilden, $SL(2)$ *Representations of Finitely Presented Groups*, Providence, RI: American Mathematical Society, 1995.

Burnside, W. F., *Theory of Groups of Finite Order*, 2nd ed., New York: Dover, 1911, reprinted 1955.

Burrow, Martin, *Representation Theory of Finite Groups*, New York: Dover, 1993.

Coleman, A. John, *Induced Representations with Applications to Sn and $GL(n)$*, Kingston, Ontario: Queens University, 1966.

Collins, Michael J., *Representations and Characters of Finite Groups*, Cambridge: Cambridge University Press, 1990.

Cotton, F. Albert, *Chemical Applications of Group Theory*, 3rd ed., New York: Wiley-Interscience, 1990.

Curtis, Charles W. and Irving Reiner, *Methods of Representation Theory, with Applications to Finite Groups and Orders*, New York: Wiley Interscience, Vol. 1, 1981, Vol. 2, 1987.

Curtis, Charles W. and Irving Reiner, *Representation Theory of Finite Groups and Associative Algebras*, New York: Wiley-Interscience, 1962.

Feit, Walter, *The Representation Theory of Finite Groups*, Amsterdam: North Holland, 1982.

Fuchs, Jürgen and Christoph Schweigert, *Symmetries, Lie Algebras, and Representations: A Graduate Course for Physicists*, Cambridge: Cambridge University Press, 1997.

Fulton, William and Joseph Harris, *Representation Theory: A First Course*, New York: Springer-Verlag, 1991.

Gorenstein, Daniel, *Finite Groups*, 2nd ed., New York: Chelsea Publishing Co., 1980.

Harter, William G., *Principles of Symmetry, Dynamics, and Spectroscopy*, New York: Wiley Interscience, 1993.

Hill, Victor E., *Groups, Representations, and Characters*, New York: Hafner Press, 1975.

International Union of Crystallography, *International Tables for X-ray Crystallography*, Birmingham, England, Kynock Press, 1952–68, v. 1. Symmetry groups, edited by N. F. M. Henry and K. Lonsdale.

Isaacs, Irving Martin, *Character Theory of Finite Groups*, New York: Dover, 1994.

James, Gordon, *Representations and Characters of Groups*, Cambridge: Cambridge University Press, 1993.

James, G. D. and Adalbert Kerber, *The Representation Theory of the Symmetric Group*, Cambridge: Cambridge University Press, 1984.

Jones, Hugh F., *Groups, Representations, and Physics*, Bristol: Institute of Physics, 1998.

Keown, R., *An Introduction to Group Representation Theory*, New York: Academic Press, 1975.

Kettle, Sidney F. A., *Symmetry and Structure: Readable Group Theory for Chemists*, 2nd ed., New York: Wiley, 1995.

Lomont, John S., *Applications of Finite Groups*, New York: Dover Constable, 1993.

Maschke, H., Über den arithmetischen Charakter der Coefficienten der Sustitutionen endlicher linearer Substitutionsgruppen, *Mathematische Annalen*, **50** (1898), 482–498.

Nagao, Hirosi and Yukio Tsushima, *Representations of Finite Groups*, Boston: Academic Press, 1989.

Navarro, G., *Characters and Blocks of Finite Groups*, Cambridge: Cambridge University Press, 1998.

Parshall, Brian and Jian-pan Wang, *Quantum Linear Groups*, Providence, RI: American Mathematical Society, 1991.

Simon, Barry, *Representations of Finite and Compact Groups*, Providence, RI: American Mathematical Society, 1996.

Stancu, Fl., *Group Theory in Subnuclear Physics*, Oxford: Clarendon Press, 1996.

Tung, Wu-Ki, *Group Theory in Physics*, Philadelphia: World Scientific, 1985.

Wielandt, Helmut, Zum Satz von Sylow, *Mathematische Zeitschrift* **60** (1954), 407–408.

Young, A., *Proceedings of the London Mathematical Society* (1), **33** (1900), 361, cited in Wehl, Hermann, *The Classical Groups*, Princeton, NJ: Princeton University Press, 1952, p. 293.

Note. Proofs of the theorems that are not proved in this text may be found in the books by Curtis and Reiner, listed above.

Index of Symbols

*Page reference is the first occurrence of the symbol.

Index

A

abelian group 15, 23
abstract group 6, 44
action of group on set 21, 37, 62, 97
alternating group 70, 76, 150, 153, 161, 194
ammonia molecule 169
associativity 11, 12
automorphism 44
automorphism group 45

B

basic primitive translation 211
benzene molecule 170
bilinear form 144
Burnside counting theorem 165

C

cancellation 18
canonical homomorphism 54
center of group 40, 60, 154
centralizer 38
character 131
character of inverse 138
character table 141, 145
characteristic roots 134, 136
characteristic subgroup 154
class function 133
closure 12
column product in character table 147
commutative group, *see* abelian group
commutator 127
commutator subgroup, *see* derived group
completely reducible representation 111
complex conjugate 137
complext norm 137
conjugate character 174
conjugate class 38, 139
conjugate, complex 137

conjugate of element 37
conjugate of subgroup 38, 47
conjugate representation 174
conjugation 37
constituent 157
coset 23
coset representative 23
cube, group of 27, 90, 159, 207
cycle 4
cyclic boundary condition 213
cyclic group 7, 13, 17, 19, 35, 119, 123

D

decomposable representation 106, 108, 111
degree of representation 85
degrees of irreducible characters 147, 180, 193
derived group 127, 154
dihedral group 8, 25, 29, 36
dimension of representation 85
direct product 120, 123

E

equivalent representations 92, 133
even permutation 70
extension problem 217
external direct product 121

F

factor group 54, 126
factor set 53
faithful representation 86
finite abelian groups, main theorem on 123
fixed point of permutation 176
formal product 51
Frobenius reciprocity theorem 191

237